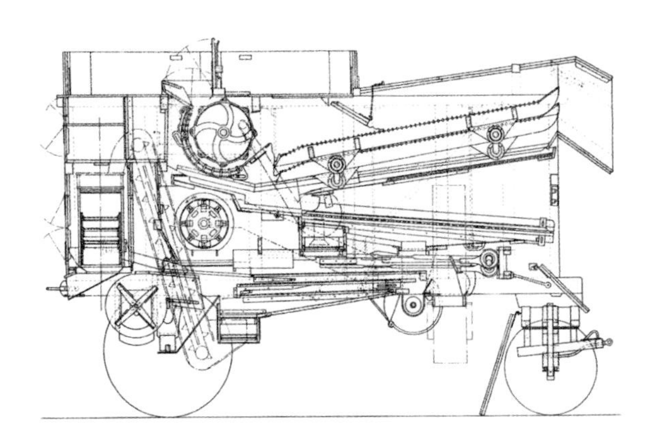

THRESHING MACHINES

Frontispiece: John Thomas' 1947 Garvie Thresher

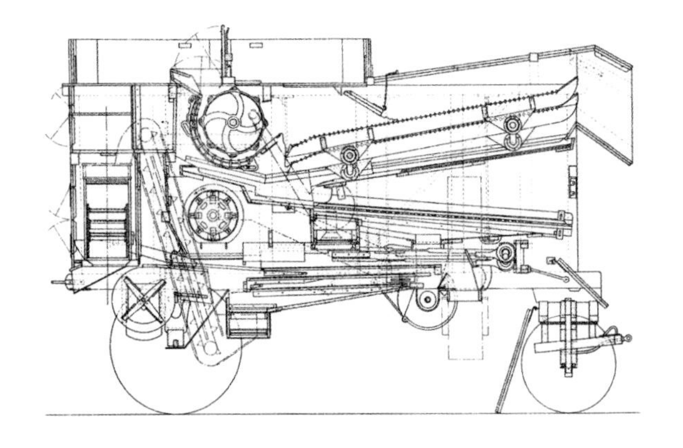

THRESHING MACHINES

Trevor Gregory

First Published 2011

ISBN: 978-1-904686-21-7

A catalogue record for this book is
available from the British Library

Published by

Japonica Press
Low Green Farm, Hutton, Driffield,
East Yorkshire,
United Kingdom, YO25 9PX

Dedication

This book is dedicated to those two great Australian Bibliophiles:

Emma Chisit
and
Emma Charthay

The Author

Trevor Gregory is proud to call himself a Cornishman who has always been interested in all things mechanical. Educated at Humphry Davy Grammar School, Penzance, Trevor left Cornwall to study Architecture at Portsmouth Polytechnic. Upon leaving college, he joined Cornwall County Council's Architects' Department and became the Council's Radiological Protection Supervisor. He has lectured and presented technical papers as far afield as Rimini and Montreal and became Radon Mitigation Consultant to the Irish Government's Board of Education. In 2005, ill health forced early retirement.

Outside work, Trevor competed (unsuccessfully) in motor cycle grasstrack racing and later (just as unsuccessfully) in motor cycle trials. Later he competed (marginally more successfully) in classic car trials. Trevor has a great interest in 20th century classical music and has played jazz guitar semi-professionally for many years. He is fluent in French and is currently attempting to master Spanish.

Interested in oenology and gastronomy (to which the waistline is testament), he has a pathological hatred of gratuitous exercise.

He has been able to use his expertise in draughtsmanship and photography in the production of this, his first book.

Contents

Introduction

An English author was once performing a signing in a Sydney bookshop. He would write "To Sheila Smith with Best Wishes from etc etc". A lady passed him a copy of his book and said "Emma Chisit" So the author dutifully wrote "To Emma Chisit with Best Wishes from etc etc". "NO, NO!" She shrieked "EMMA CHISIT". Eventually the author realised that the lady was speaking in phonetic Australian, or as they say, "Strine". She was really saying "How much is it?" It is to this perceptive lady and to her best friend, Emma Charthay, that this volume is respectfully dedicated.

A visitor from another world might find earth's preoccupation with cereal production all a little odd. On a planet where fruit can be picked and eaten directly off a tree or bush, or a root vegetable can be dug up and boiled then eaten, cereals seem a much more daunting option. Early man would have had to have searched out suitable small wild grasses, tear them up or cut them down, separate the few small seed heads, remove their husks, grind them down, mix them with water, and then bake them. Perhaps the only saving grace is that grass is one of the largest and varied families in the plant kingdom. Some species or other will flourish in virtually any location, be it arid or swampy, arctic or tropical, exposed or sheltered.

The era covered by this book is very much an imperial one; not just pounds, shillings and pence, feet and inches, pecks and bushels… but one where large areas of the globe were coloured pink and the sun never set on them. For this reason, all the units used in this book will be the original imperial ones.

As well as unfamiliar units of measurement, there exists a whole group of specialist vocabularies. Just as Eskimos are supposed to have 9 words for snow, so farmers seem to have a whole range of words just for the various pieces of cereal crops. And to make matters worse, manufacturers have their own agenda, and are less than consistent in the use of names for the various parts of their products.

The first machines were called either thrashers or separators which meant that they thrashed the crop, and separated the grain from the stalks. Dividing the grain from all the other material, especially from the smaller pieces of straw and husks, had been performed by blowing this lighter material away in a totally independent process called winnowing. Thus early machines only threshed, the winnowing being carried out manually or with a separate winnowing machine. When the two processes were later combined, the two-stage machine was still known as a thresher. The final change came when techniques for dressing and grading the grain were added at the end of the process. For the sake of clarity, it is a pity that no-one thought to find other words for these other machines. So regrettably, in this book the word thresher has had to be applied to all types.

Go to a steam rally today and visit the bookstalls. How many books are there about traction engines?. How many books are there on threshing machines? It would be difficult to write about threshing machines without making some reference to the motive power that they relied on. It's all too easy to yield to the pressure of writing yet another history of the traction engine. It feels like trying write the biography of an unknown man with a very famous brother. Perhaps the situation is best described by the person when talking about preservation, who said "traction engines without threshing machines are like cups without saucers".

As this is the author's first attempt at a full-sized book, some inspiration has been needed. In terms of layout, my college history lecturer's advice to students formatting their dissertations was: "Say what you're going to say, say it, then say what you've said".
In terms of explaining exactly what happens when the crop passes through the machine, the author is reminded of the building services lecturer at his School of Architecture who used to say: "When designing sewerage systems, imagine that you're a turd... He would also add "Some of you may find this easier than others".

Finally there are the immortal words: advice is something you ask for when you know the answer... and wish you didn't.

Chapter One

BEFORE THE MACHINE

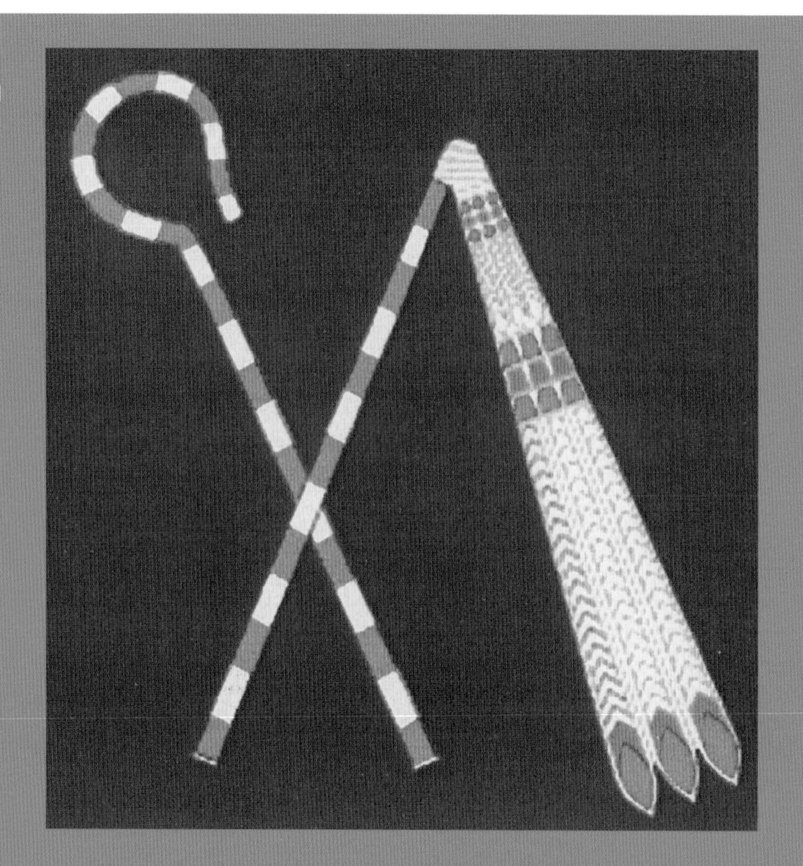

Egyptian Pharoah's Flail & Shepherds' Crook

Threshing, also spelt thrashing or even tireshing is defined as:

"The process by which the grain or seed of cultivated plants is separated from the husk or pod which contains it".

The words themselves are derived from verbs in Anglo-Saxon languages meaning to thresh. Old English is *terscan* or *erscan*, German is *dreschen*, Dutch is *dorschen*. Interestingly, these words do not have other meanings which goes to show how important this activity really is, having a word all of its own.

Before the arrival of machinery specifically designed for the purpose, the separation of individual grains from chaff and stalks was a very labour-intensive and time-consuming procedure requiring great care if the maximum amount of clean grain was to be extracted. The process involves 2 distinct steps:

1. The mechanical separation of the kernels from the plant heads or husks.
2. The sorting of the kernels from the remaining extraneous material.

Before the advent of machinery, farmers relied on humans, and in some instances, livestock, to provide the necessary labour. It is probable that in the earliest times, the little grain that was grown was shelled by hand, but as the quantity increased, the bundles or sheaves would have been beaten upon the ground. Alternatively the grain could be struck with a stick.

The Flail

An implement consisting of handle with a free swinging stick at the end; used in manual threshing. From the Latin flagellum meaning whip or scourge

In Hebrew the word for this stick is an *aba*. Doubtless the flail was evolved from the early method of using the stick. It seems to have been the thrashing implement in general use in all Northern European countries, and was still the chief means of thrashing grain as late as the 1860's. The 2-piece tool we think of today is believed to have originated in Gaul and one of the earliest references to it was made by St Jerome in AD 420. Once the crop had dried, the individual bundles or sheaves could be carried to a sheltered location, to be spread out on a flat, clean surface in preparation for separation. Labourers then struck the stalks of wheat with the flails, knocking the kernels out of the heads. Often the threshing didn't start until two months or more after harvest. This is because grains can be flailed most easily not only when the crop is fully dry, but also it is best performed on sharp cold days.

The flail itself consists of two pieces of wood; the handstaff or helve, and the beater. The two are fastened together loosely at one end by a thong. Rawhide or eelskin is often used to make what needs to be a very durable joint. The handstaff is a light rod of ash slightly increasing in diameter at the farther end to allow for the hole for the thong to bind it to the beater. Ash is chosen for its natural springiness and interestingly, was used in the construction of certain parts of modern threshing machines where this springiness was required. The handstaff is about 5' 0" in length and this enables the operator to stand upright while working, which makes it much easier on the back in view of the time likely to be spent at what is a very lengthy, not to say repetitive task. The beater is a wooden rod about 2' 6" long. Often this too is made of ash, though a more compact wood such as thorn is sometimes used. The wood for this part of the flail is chosen for its resistance to splitting. The beater also has a hole at one end for the thong to bind it to the handstaff. The shape of the beater is cylindrical, of about 1¼" diameter and constructed so that it is the end grain of the wood that receives the force of the blows. An average speed of work would be in the order of 30 to 40 blows or strokes per minute. In some places, those using flails wore shoes made from old hats so as not to bruise the grain.

Fig 1: Medieval Psalter - Threshing by Flail

In one day, an average labourer could be expected flail the following quantities:

CROP	BUSHELS PER DAY
Oats	30
Buckwheat	20
Beans	20
Barley	16
Rye	8
Wheat	7 to 10

In John Morton's 1830 publication, "Handbook of Farm Labour", a calculation is presented where 1400 ft/lbs of energy is required to thresh 4 quarters or 18 cwt of wheat using a flail. This equates to 7 man days for each acre of the crop.

Fig 2: Ransomes Catalogue - Threshing by Flail

As threshing continues, the amount of grain released decreases. By the law of diminishing returns, farmers often felt obliged to spend more time supervising their labourers towards the end of threshing sessions.

Despite widespread mechanisation, the flail is still in use for special purposes such as flower seeds, and also where the quantity grown is so small as to render it not worthwhile to use a machine.

The flail was well-known in Japan from the earliest times, but here it was often used in conjunction with a stripper, an implement fashioned very much like a large comb, with the teeth made of hardwood and pointing upwards. After being reaped, the crop was brought to this fixture and combed through by hand, the heads being drawn off and afterwards thrashed on the thrashing floor by the flail. Today just such a comb type implement, known as a heckle, is still used for removing the boils or heads from flax.

Livestock Treading

The action of the hooves of livestock treading on piles of grain can also used as a method of separating grain. As quantities increased, it was the practice of the ancient Egyptians and Israelites to spread out loosened sheaves on a circular enclosure of hard ground 40' to 100' in diameter, and drive oxen, sheep or other animals round and round over it so as to tread out the grain. In Hebrew *dush* means to trample hard, the level floor being a goren. It had been long recognised that care needed to be taken when spreading out the sheaves. A Roman writer described in some detail how corn awaiting threshing should be heaped to avoid damage as it lay on he floor.

In colonial America, George Washington experimented with a derivation of this method of threshing. On his farm he built a round barn, the first of its type in America, and stacked it with wheat up to 3 feet high. Workers would then take specially-trained horses and have them run round and round on the wheat. The grain would be dislodged from the stalks by the action of the horses' hooves and then fall through special gaps in the planking of the floor to a storage chamber below.

Fig 3: Animal Treading - New England Round Barn

Despite the greater efficiency when compared with the use of the flail, it was always recognised that livestock treading treated the crop far more roughly. For this reason, where a number of different threshing processes were use in the same area, barley, oats, and buckwheat would generally be trodden under animal hooves, whereas the less robust wheat and rye would be flailed.

Sledges, Boards, and Rollers

In some areas, the practice of livestock treading was partially superseded by using the animals to pull implements over the grains. There exists a clay tablet found in Iruk, southern Iraq, dating from 3,000 BC which depicts a simple threshing sledge. In Jewish culture, these implements were known as the *Aru* and the *Morag* (mow-rag) both of which, to judge from their modern representatives, were heavy wooden drags, sometimes weighted with large stones and/or with the driver. One ancient Hebrew text complains that " The driver today not infrequently reposes at full length upon the drag, and even slumbers, while the docile oxen follow their monotonous round over the straw". If the oxen had a mind of their own, one can imagine a driver waking up miles from home. The humane legislation of the Pentateuch mentioned in Deuteronomy Chapter 25 forbids the muzzling of the oxen while treading out the corn and the Talmud similarly "enjoins that they be blindfolded as a safeguard against dizziness". One feels that the RSPCA would heartily approve.

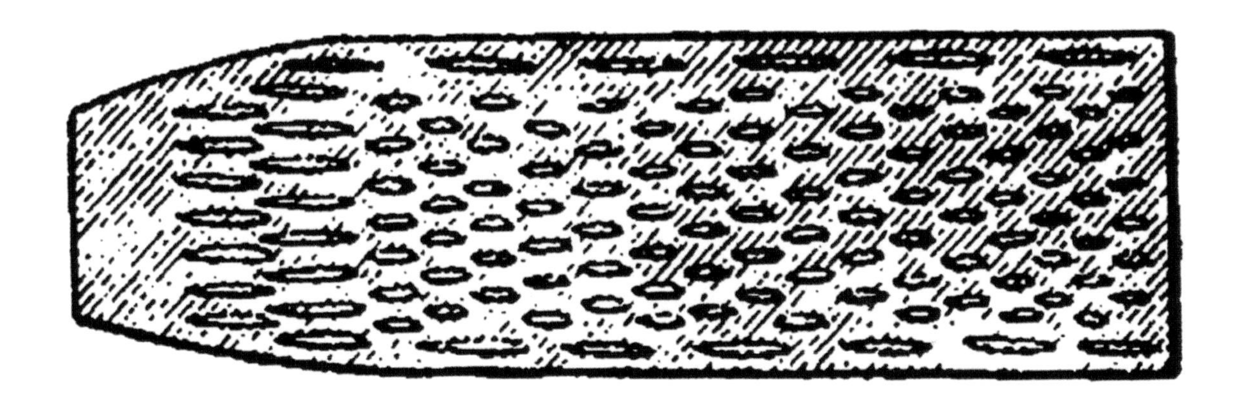

Fig 4: Sledge

The Egyptian equivalent of the Jewish *morag* was the noreg or "corn-drag" consisting of a wooden frame with 3 cross members. The undersides of these drags were sometimes modified either with revolving metal disks, or, more commonly, with projecting teeth either of stone or basalt. These were the size of a walnut inserted tightly into holes in the drag, so as to protrude a couple of inches. The *norag* of northern Spain uses sharp flakes of flint driven into a framework of thick boarding. Just as concerns were expressed regarding the rough treatment of grain when not threshed by hand, there are biblical references to similar damage done to straw caused when using animals and drags. Presumably this was seen as a serious problem where thatching was common practice with long unbroken straw at a premium. A report claims that over 2 million examples of the Turkish *doven* or sled were still in use in the 1960's. They were generally used on a circle of corn of 40' in diameter and 12" deep. This method resulted in grain losses in the order of 30%. The straw was also badly broken but was still suitable for feeding directly to livestock. It could also be used as bedding, but not for thatching. The simple sledge with a ridged or grooved bottom was used by the ancient Greeks and similar methods are still employed in many countries today.

Fig 5: Sledge - Hebrew Mow-rag

The Iranian *gharji* featured two spiked rollers and the later Egyptian *charatz* or "threshing board" consisted of a heavy frame mounted with three or more rollers, but only sometimes fitted with spikes. These rollers were drawn over the spread-out corn by pairs of oxen and records exist showing this being in use prior to 1,500 B.C. The rollers can be cylindrical or tapering.

The next stage of development was to fasten one end of the roller to an upright shaft in the centre of the thrashing floor and pull it around from by its outer end. This arrangement is still used in Italy and is a direct descendant of the Roman roller sledge known as the Tribulum. This is the instrument that is the origin of the modern word tribulation meaning literally to be ground down. The rollers could be plain, fluted, or spiked, and the spiked or "porcupine" pattern wooden items were used in the USA as late as the 18th century. Some examples were over 12' long with as many as 250 individual wooden spikes.

Winnowing

The traditional threshing floor was often situated on an elevated piece of ground so that as the straw was removed, wind could blow away the chaff leaving the corn. For grain to command top price when sold for grinding for flour, it is imperative that it contains as little chaff and weed seeds as possible. This same principle also holds true for grain intended for planting as the next year's crop. Fewer weed seeds mean a better overall yield.

After the grain had been beaten out by the flail or ground out by other means, the straw would be carefully raked away and the corn and chaff collected for separation by winnowing.

The material would be placed in a winnowing basket, on a tray, or on a blanket. The material would then be tossed into the air, where currents of air would carry the lighter chaff away, allowing the grain to fall back into the receptacle. As this method of winnowing takes place, the best grain falls more vertically whilst the lighter grains are carried some distance before falling thus allowing a very rough-and-ready grading of the crop to take place. Another alternative is to use a purpose-designed fanning shovel. These are of a distinctive pattern with a very long and narrow blade.

The currents of air were usually provided by the wind, but waiting for this could pose problems. The process could still be performed when there was no wind by pouring a stream of the mixture from a vessel in front of a fan. In Jewish culture, the larger amount of worthless straw torn into short lengths by the weighted teeth of the *morag* made winnowing a more necessary operation, as well as a more tedious one. The mass of mingled grain, chaff, and short straw could also be tossed into the air using a shovel. In Hebrew, this was the *mizreh* from the word *zarah* meaning to scatter and the *ra-at* connected with *rua* or wind, and implements under these names are still used in Palestine to-day. The lighter material, chaff and tail corn, would be used for animal feed or *teben*, whilst the heavier best corn intended for human consumption would fall at the winnower's feet. This grain would be further dressed by being shaken through a sieve or *kebarah*. The mesh of the Palestinian sieve of today is still made of slips of dried camel-hide, and is fine enough to allow the threshed grains to pass through whilst retaining the unthreshed ears. These are then collected and returned to the threshing-floor to be processed a second time.

In areas of higher rainfall, threshing took place indoors. Barns were traditionally constructed with their axis in line with the direction of the prevailing wind. Built with large doors at either end, this allowed the wind to blow right through the barn and across the threshing floor and enabled winnowing to continue well into the autumn. Writing about primitive Scottish winnowing practices, a Mr Ritchie describes a special hand-tool used in these barns consisting of a sheepskin with holes cut in it, stretched on a 2" diameter wooden hoop.

Threshing in the West

Threshing Anecdotes

Farmers are not generally known for their generosity but tradition dictated that it was the duty of their wives to prepare "croust" for the whole threshing team. However they often received instructions from their husbands.

The tea was always to be dispensed first. This was from the large urn, in the style of a Russian Samovar, into the large thick white mugs. This was intended to blunt the appetite before the contents of the wicker baskets with their gingham tea-towel covers were revealed. In Cornwall, a common resident of the wicker basket was the pasty. While some would have contained some meat, there was usually more vegetable material what old Cornish people refer to as "taties and point" meaning that you could see the veg but you had to point to where the meat was. Often the pasties contained jam which was intended to reduce victualling costs still further.

On one occasion, the threshing had to be stopped just before lunch because of torrential rain. The farmer's wife had already prepared the "croust", but when it appeared, all the pasties had been cut in half. This was to reflect the fact that only half a day's work had been completed

Chapter Two

CROPS

Corn Dollies made from Oats

"History... knows the names of the king's bastards but cannot tell us the origin of wheat" J H C Fabre 19th century French Botanist and Entomologist

Cereals

The word cereal is derived from Ceres, the Roman goddess of harvest and agriculture. All cereal crops are really species of grasses cultivated for their edible grains or seeds. These seeds are technically a type of fruit called a caryopsis. Cereal grains are grown in greater quantities worldwide than any other type of crop and provide more food energy to the human race than anything else. Indeed over 80% of the world's food is provided by less than a dozen plant species. In some third world countries, cereals constitute practically the entire diet of large sections of the population. In developed nations, cereal consumption is more moderate, but still very substantial. Wheat, together with maize,and rice accounts for 87% of all grain production, worldwide and 43% of all food calories.

Whilst each individual species has its own peculiarities, the cultivation of all cereals crops is similar. All are annual plants; consequently one planting yields one harvest. Wheat, barley, oats, rye, and the more recent hybrid, triticale, are referred to as the "cool-season" cereals. These are hardy plants that grow well in moderate weather but cease to grow in temperatures much greater than 30 °C, although there is some variation depending on species and variety. Whilst wheat is the most popular, barley and rye are the hardiest cereals, able to survive the winter in the sub-arctic and Siberia. Many cool-season cereals are grown in the tropics, but the majority would only be grown in cooler highland areas. Here it may even be possible to grow more than one crop a year. Once the cereal plants have grown their seeds, they have completed their life cycle, they die and become brown and dry. As soon as this happens, harvesting can begin. In developed countries, cereal crops are universally machine-harvested, now using a combine harvester, which cuts, threshes, and winnows the grain during a single pass across the field but, of course, this book is looking at a previous era.

Cereal grains supply most of their food energy as starch. They are also a significant source of protein, though the amino acid balance is not optimal. Whole grains are good sources of dietary fibre, essential fatty acids, and other important nutrients. Oats are rolled, ground, or cut into bits, a form known as steel-cut oats, and cooked into porridge. Most other cereals are milled, or ground, into flour or meal. The outer layers of bran and germ are removed which lessens the nutritional value but makes the grain more resistant to degradation because the outer layers of the grains are rich in rancidity-prone fats. Health-conscious people tend to prefer whole grains, which are not milled, which makes the grain more appealing to many palates, but over-consumption of milled cereals is often blamed for obesity. However, milled grains do keep better. The waste from milling is sometimes mixed into a prepared animal feed. Once milled and ground, the resulting flour can be made into bread, pasta, desserts, dumplings, and many other products. Cereals are the main source of energy providing about 350 kcal per 100 grams, but cereal proteins are typically poor in nutritive quality, being deficient in the essential amino acid, lysine.

The yields quoted later in the text are generally early 20th century figures. This is in order to reflect the time when threshing machines were most commonly in use.

Unlike 21st century practice, straw was very highly-valued commodity and was often what prompted the occasional threshing session by small farmers, rather than the grain.

Table 1: Bushel Weight of Grains

BUSHEL WEIGHT ETC. OF COMMON CROPS		
	Wt. per bushel (lb.)	Cu. ft. per ton
Wheat	63	46
Oats	42	70
Barley	56	51
Rye	57	51
Peas	63	46
Beans	66	43
Potatoes	50	54

Cereal morphology - a little Botany

A threshing machine designer studying cereal botany must surely be like an executioner studying human anatomy

Ignoring germination and root development, the structure of adult cereals is as follows:

The vertical stem or *culm* consists of a hollow cylinder punctuated at the top and at a number of points along its length by solid *nodes*. The lengths of stem between the nodes are known, somewhat predictably as *internodes*. All growth takes place from buds located on the nodes. Often the nodes are covered with hair. When this does not occur, these nodes are described as *glabrous*. Vertical growth, essentially elongation of the internodes, is known as *jointing*. Some wheats can reach as much as 7' in height, although cultivated versions are more like 4'. Semi-dwarf *cultivars* of wheat can be as short as 3'. A single leaf grows from each stem node. These are arranged alternately at 180 degrees to each other. The leaves begin to grow enclosing the stem in a leaf *sheath* up as far as the next node before increasing in length and moving away from the stem at 90 degrees to become a *flag* leaf. A smaller secondary stem can appear from a bud on the node at the bottom of the stem. This is known as a *tiller* and is of the same construction as the main stem. Flower development or *inflorescence* occurs at the topmost node or *crown* in an area called the *meristematic region*. The flowers develop along a flattened central stem called a *rachis* in a layout whose botanical name is a *spike*. On some varieties, the layout sub-divides to create smaller *spikelets* carried on smaller stems called *ratchillas*. Each flower forms as a *glume* and is potentially a grain *kernel*. Different varieties can be identified by glume size and shape. The particular features to be noted are the *keel*, the *beak*, and the *shoulder*. Both male and female flowers occur on the same plant. The female flowers with their ovaries are known as *palea*, and the males with their stamens, *lemmae*. It is the male flowers that can grow long to form *awns*. Empty glumes are sterile. The grains, when they form, are composed of *endosperm* enclosed in a series of sheathing layers.

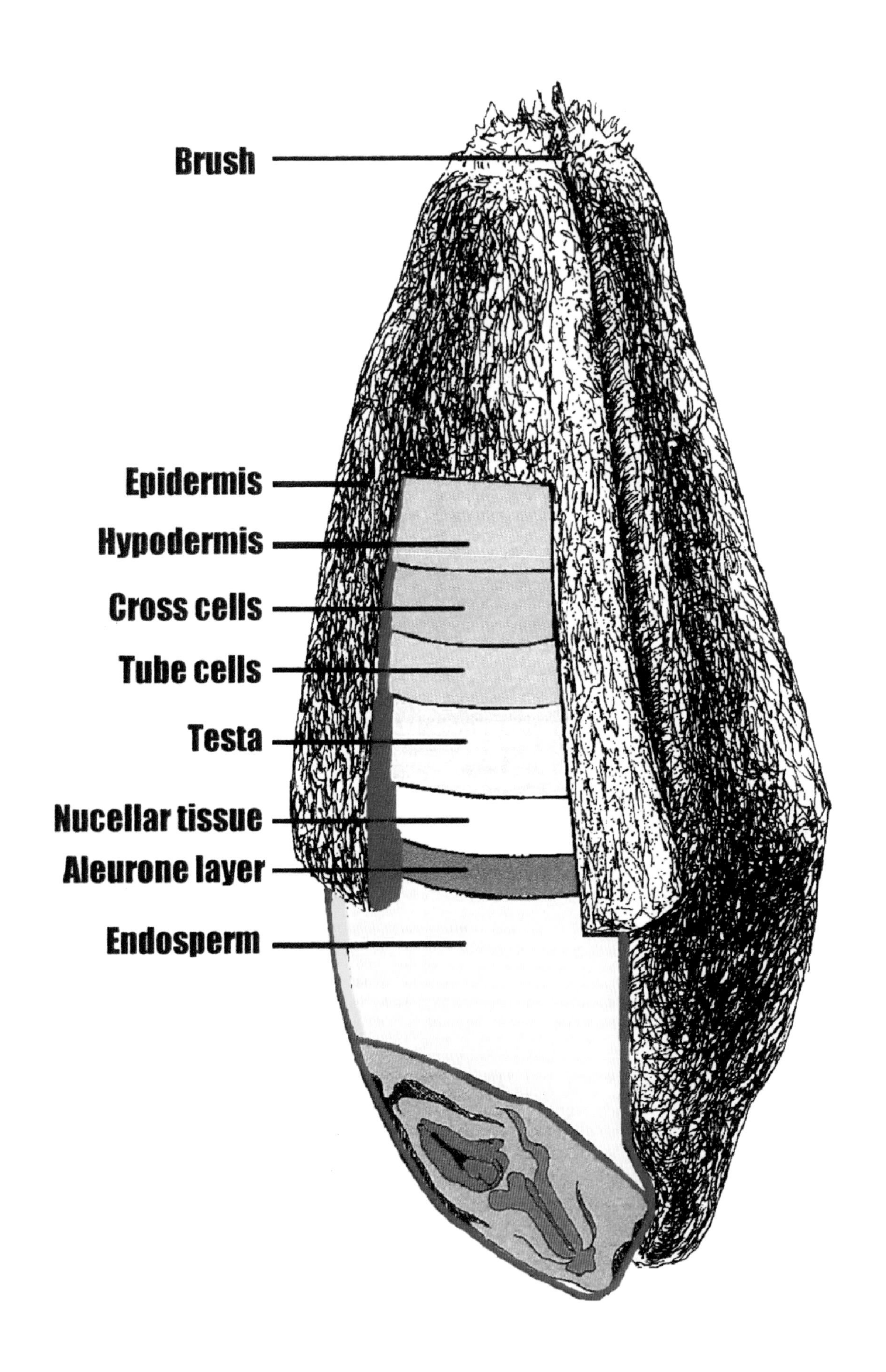

Fig 6: Barley Kernel

Harvest Options

All cereals can be grown as winter or summer crops. Winter barley sown in October / November will be ready to harvest in July, whereas sowing in March / April means harvest in September. Prices do not differ between the two as they do in "early" potatoes for example. Because of the extra time spent in the ground, the winter crop will produce a yield of around a ton per acre more. It is also more hardy and resilient. This is particularly so if it is rolled at the 3 to 4 leaf stage of development. The earlier harvest date also carries the advantage of more daylight providing a correspondingly longer working day. The major advantage of a spring sowing is that it can take place immediately after another harvest such as cabbage or cauliflower thus allowing two crops a year.

Optimum moisture content for grain at harvest is around 16%. Below this when the grain is "dead ripe" the grain tends to shatter or fall off the stalks, particularly when struck by the reel of the binder or combine. But delay in the harvest can result in losses of up to 1 cwt per acre per day and can even allow sprouting in the ear if too wet.

Interestingly in the context of this book, a binder can be used 3 weeks before a crop would be suitable for combining. The crop can be allowed to dry out naturally in the rick where it benefits from a level of air circulation that would not be possible on the ground alone. The down side of this practice occurs when sheaves still in the field are soaked by rain. This necessitates turning them to enable them to dry, an unpleasant and time-consuming process. Some older varieties were even thought of as being unsuitable for tying using a binder so required laborious hand-tying.

Fig 7: PTO-driven Binder

Feed Options

Fig 8: The Digestive system of the Horse

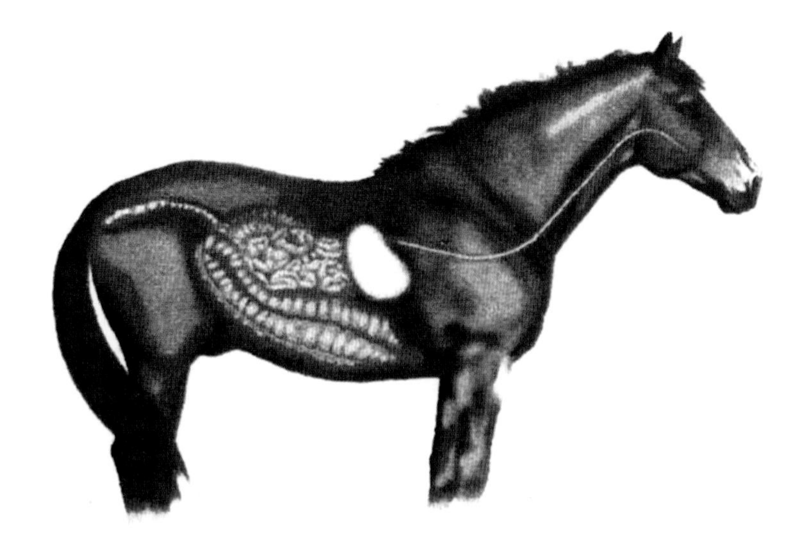

During the era of the threshing machine, very much more of the cereal crop was grown as animal feed. This was due in large measure to the number of horses working the land. It has been said that up to one third of the total acreage of a farm was growing crops to feed the horses working it.

A horse's stomach is the size and shape of a rugby ball. In the wild, the horse would be browsing continuously on relatively low-protein food. From a purely working perspective, when domesticated, it would be best if small amounts of high-protein food could be taken quickly at infrequent intervals. This is just not viable and feeding has to be arranged to replicate as far as is reasonable, their natural environment. This means making fodder available over as long a period as possible. Also horses are not able to operate on a high-protein low-quantity diet. The richness of grain alone can result in digestive problems such as colic. Barley in particular is often referred to as a "hot" feed and in addition to digestive problems, can also cause "hot hooves" which are highly liable to infection. The problem is lessened if some or all the barley is replaced by oats. As a digression, the final treatment for grain before being used as feed varies depending on the animal for which it is destined. Sheep are able to digest whole grain, whereas cattle require barley to be rolled, and pigs take their grain milled as flour.

The key to successful feeding of horses is to provide plenty of roughage in the form of hay or chaff. Perversely, barley straw cut as chaff is given to cattle that become "loose". Early and late feeding times plus making roughage available as much of the time as possible are the goals. A good example of this would be that of the ploughman who stops for a snack and gives his horse the nosebag at the same time. A First World War directive to the cavalry makes the following recommendations for feeding:

- Horses should eat as early and as late as possible.
- At 7.00am, they should be given 4lb of oats plus a handful of chaff.
- This should be repeated at 11.00am.
- They should be able to take Hay from nets from 7.00 am onwards.
- A nosebag containing 2½ lb of chaff should be available whenever possible.

The above represents a daily allowance of 8 lb of oats and 3 lb of chaff. A recent publication on the subject of feeding heavy horses recommends that: *"In addition to grass in summer and hay in winter, the horse should receive 6 lb of rolled barley plus chaff twice a day"*. A heavy horse weighing 1 ton is thought to need 2 to 3 acres to graze on.

As an aside, hay can be made more digestible if it is subjected to bruising, a practice more common in warmer climates where the crop is dryer. This is often achieved using a mower fitted with a "conditioner", or a separate machine known as a "crimper". These both consist of a pair of fluted rollers very similar to those used to feed early threshing machines, and which were finally abandoned because they caused so much damage to the straw.

Fig 9: Horse-drawn Binder

Fig 10: Hay Roller-crusher or Crimper (New Holland)

Barley

Fig 11: Barley

Kingdom: *Plantae* Division: *Magnoliophyta* Class: *Liliopsida*
Order: *Poales* Family: *Poaceae* Genus: *Hordeum*
Species: ***Hordeum vulgare*** Varieties: ***Sprat Archer, Abed Maja, Craig's Triumph***

Barley has three main uses; animal feed, human food, and brewing. It is a member of the grass family Poaceae. Different varieties of barley are identified by the number of kernel rows in the head. Three forms are cultivated: two-row barley traditionally known as *Hordeum distichum*. four-row or *Hordeum tetrastichum*. six-row barley or *Hordeum vulgare*.

Along the rachis are arranged alternately, up to 40 groups of 3 florets. In two-row barley only one of these florets is fertile whereas in the four-row and six-row forms, all three are fertile. Two-row barley is the oldest form and shares this configuration with wild barley. Two-row barley has a lower protein content than six-row barley and thus a lower enzyme content. High protein barley is best suited for animal feed or malt that can be used to make beers with a large adjunct content. Two-row barley, the most common variety, is traditionally used in English ale-style beers while six-row barley with 3 seeds on each side is traditional in German and American lager-style beers. The four-row version is unsuitable for brewing. The stems can grow up to 5' in height, although cultivated varieties are unlikely to exceed 4'.

Next to wheat, barley is the most valuable grain and grows especially well on light and sharp soils. It is more tolerant of soil salinity than wheat, which might explain the increase of barley cultivation the fertile crescent area of Mesopotamia from the 2000 BC onwards. Barley can still thrive in conditions that are too cold even for rye but it is a tender grain and easily hurt in any of the stages of its growth, particularly at seed time. Barley is raised at greater expense than wheat, so its cultivation is often reserved for rich and genial soils, and where climate will not allow wheat to grow easily. The quantity sown differs according to a number of circumstances as well as the quality of the soil. Upon very rich lands 100 lbs per acre are typically sown, 140 lbs is very common, and upon poor land, even more is required. The seed is generally drilled in rows 6" to 7" apart. The correct quantity will produce a crop with few offsets, growing and ripening evenly, to deliver a grain that is uniformly good.

Barley is best harvested when the kernels are not quite completely dry. In harvesting and threshing, greater pains and attention are required to ensure success than in the case of all other grains. The threshing process is difficult because the awn generally adheres to the grain, and makes separation from the straw a more troublesome task.

Barley grains with their hulls still on are called covered barley. Once the grain has had the inedible hull removed, it is called hulled barley or pot barley. At this stage, the grain still has its bran and germ, which are high in nutritional value. Hulled barley is considered a whole grain, and is a popular health food. Pearl or pearled barley is hulled barley which has been processed further by "polishing" in a revolving drum to remove the bran and the germ. It may be processed into a variety of Barley products, including flakes similar to oatmeal and is often used as a thickening agent for soups. By-products of pearl barley are animal feeds and flour. The flour is used to make bread and baby cereal.

Barley can also be malted for use in the production of alcoholic beverages, either as beer or in distilled form as whisky. These uses attracts the highest prices, but this has to be balanced against greater demands in terms of quality control. Barley to be used for malting is often preferred if some of the beards or bristles, known as awns, are left on the grain as this aids the fermentation process. Maltsters are known to say of different barley varieties "Long awns make better malting, short awns denote a more steely grain". Classification of barley varieties began before the end of the 17th century, and has always concentrated on making the distinction between those suitable for malting and those for food purposes.

Malting requires the grain to be steeped in cool water before being spread across drying floors where the grains sprout. The malting process releases the enzymes known as amylases which convert the natural starches into sugars. Non-alcoholic drinks such as barley water and mugicha are also made using unhulled barley. Like all cereal crops, barley is affected by smut particularly by a variety known as covered smut.

Barley straw cut as chaff is used as a major source of fodder, its fibres being longer than those of other cereals such as oats for example.

Wheat

Kingdom: *Plantae* Division: *Magnoliophyta* Class: *Liliopsida*
Order: *Poales* Family: *Poaceae* Subfamily: *Pooideae*
Tribe: *Triticeae* Genus: *Triticum*
Varieties: ***Rampton Rivet, Petit Quin Quin, April Bearded***

Wheat is a grass that is cultivated worldwide. Globally, it is the most important human food grain and ranks second in total production as a cereal crop only to maize; the third being rice.

Fig 12: Wheat

Wheat can be planted as a forage crop for livestock and the straw can be used as animal fodder, or for thatched roofing. There are two principal varieties: common wheat or Bread Wheat - *(Triticum aestivum)* A hexaploid species that is the most widely cultivated in the world, and durum - *(Tritcum durum)* the only tetraploid form of wheat widely used today, second only to common wheat. The stems grow up to 7 feet in height although cultivated varieties are unlikely to exceed 5 feet.

It is the first cereal known to have been domesticated starting around 10,000 years ago with wild einkorn *(Triticum monococcum)* and emmer in the fertile crescent region of Mesopotamia. Cultivation and repeated harvesting and sowing of the grains of wild grasses led to the selection of mutant forms. These featured tough ears which remained attached to the ear during the harvest process, as well as a larger grain size. A side effect of this selection process was of the loss of its natural seed dispersal mechanism which prevents domesticated wheats from surviving in the wild. Selective breeding has softened the tough glumes that tightly enclose the grains whilst encouraging a semi-brittle rachis that breaks easily on threshing. The result is that when threshed, the wheat ear easily breaks up into spikelets.

To make the grain useful, further processing, such as milling or pounding, is needed to remove the hulls or husks. Hulled wheats are often stored as spikelets because the toughened glumes give good protection against the pests that can attack stored grain.

Yields of wheat increased greatly following the change from broadcasting seed to the use of seed drills that began in the 18th century. In Western Europe target wheat yields attainable are now around 65 cwt per acre. While winter wheat lies dormant during a winter freeze, wheat normally requires between 110 and 130 days between planting and harvest, depending upon climate, seed type, and soil conditions. The meïosis stage is extremely susceptible to low temperatures (under 4 °C) or high temperatures (over 25 °C). Farmers also benefit from knowing when the topmost flag leaf appears, this represents about 75% of photosynthesis reaction.

Much of the following text is taken from the Household Cyclopedia of 1881:

Wheat may be classed under two principal divisions, though each of these consists of several subdivisions. The first is composed of all the varieties of red wheat. The second division comprehends the whole varieties of white wheat, which again may be arranged under two distinct headings, namely, thick-chaffed and thin-chaffed.

Thick-chaffed wheat varieties were the most widely used before 1799, as they generally make the best quality flour, and in dry seasons, equal the yields of thin-chaffed varieties. However, thick-chaffed varieties are particularly susceptible to mildew, while thin-chaffed varieties are quite hardy and in general are more resistant to mildew. Consequently, a widespread outbreak of mildew in 1799 began a gradual decline in the popularity of thick-chaffed varieties.

Wheat is widely cultivated as a cash crop because it produces a good yield per unit area, grows well in a temperate climate even with a moderately short growing season, and yields a versatile, high-quality flour that is widely used in baking. Most breads are made with wheat flour, including many breads named for the other grains they contain like most rye and oat breads. The vast majority of the crop is used by the milling trade for flour. As well as being used for leavened, flat and steamed breads, biscuits, cakes, pasta, noodles and couscous are wheat-based. Some is used for fermentation to make alcoholic drinks in the form of beers and vodkas, and more recently, increasingly large amounts are being directed for use as a biofuel.

Wheat is affected by smut, in particular, a variety known as stinking smut. Unlike Barley and oats, wheat straw is more suitable as a bedding material being more absorbent and softer.

Oats

Kingdom: *Plantae* Division: *Magnoliophyta* Class: *Liliopsida*
Order: *Poales* Family: *Poaceae* Genus: *Avena*
Species: ***Avena sativa*** Varieties: ***Ayr Ally, Golden Rain, Glasnevin Ardri***

The Oat is a species of cereal grain, and the same name is applied to the seeds of this plant. A now obsolete Middle English name for the plant was haver, still used in most Germanic languages and surviving in modern English as the name for a livestock feed bag, the haversack.

The heads known as panicles consist of 40 to 50 individual flower clusters or spikelets each containing 2 oat seeds or groats within the husk or hull. The stems grow from 24" to 48" high.

Fig 13: Oats

Oats are cold-tolerant and are not affected by late frosts or snow. They have a lower summer heat requirement and greater tolerance of rain than other cereals and so are particularly important in areas with cool, wet summers such as those found in north-west Europe. Certain varieties have even being grown successfully in Iceland. It exists in both winter and spring sowing varieties but an early start to sowing is crucial to good yields as oats become dormant during the heat of summer. Typically about 90 lbs, or 2 bushels, are sown per acre, either by broadcasting, or by drilling in 6" rows.

Oats are much slower to mature than barley or wheat. The are often harvested when yellow and dry with a 12% to 14% moisture content. Best yields are obtained by cutting the plants at about 4" above ground and putting them into windrows with the grain all oriented the same way just before the grain is completely ripe. The windrows are then left to dry in the sun for several days. There is a Scottish expression "leave them 3 Sundays, then pick them up". Oats can also be left standing until completely ripe but this leads to greater field losses. The grain tends to fall from the heads or accidentally threshed during cutting. A good yield is typically about 24 cwt, or 100 bushels, of grain per acre plus two tons of straw.

Oats are used for food for people and as fodder for animals, especially poultry and horses. Oats have a higher food value than any other grain, and after maize, oats have the highest lipid content of any cereal at around 10% compared with only 2% to 3% for wheat and most other cereals.

Since oats are unsuitable for making bread on their own, they are often served as a porridge made from crushed or rolled oats, oatmeal, and are also baked into biscuits known as oatcakes which can have added wheat flour. They are made into cold breakfast cereals, and are an important ingredient in muesli. Oats may also be consumed raw, and some biscuits are made with raw oats. The discovery of the healthy cholesterol-lowering properties of oats has led to wider appreciation of oats as human food. Oat bran is the outer casing of the oat and its consumption is believed to lower cholesterol thereby reducing the risk of heart disease. Bodybuilders are known to eat copious amounts of oats as a major source of carbohydrate.

Oats are also occasionally used in Britain for brewing beer, oatmeal stout being one variety brewed using a percentage of oats for the wort. In Scotland, oats are held in high esteem as a mainstay of the national diet. Traditionally the English have said that "Oats are only fit to be fed to horses and Scotsmen", to which the Scots reply "England has the finest horses, and Scotland the finest men". Samuel Johnson notoriously defined *oats* in his *Dictionary* as "A grain, which in England is generally given to horses, but in Scotland supports the people". Oats also have non-food uses and the extract can be used in bath and skin products.

Like all other cereals, oats are affected by smut, in particular, a variety known somewhat predictably known as oat smut.

Oat straw has short fibres and is used as animal bedding and sometimes as animal feed. It can be used for making corn dollies and is the favoured filling for home-made lace pillows. Even more than barley, oat straw cut as chaff is a major source of fodder. In choosing which variety is to be grown, greater account is taken regarding straw production than with other cereals.

Rye

Kingdom: *Plantae*	Division: *Magnoliophyta*	Class: *Liliopsida*
Order: *Poales*	Family: *Poaceae*	Subfamily: *Pooideae*
Tribe: *Triticeae*	Genus: *Secale*	Species: ***Secale cereale***

Rye is a grass grown extensively both as a grain and forage crop. It is a member of the wheat tribe, triticeae, and is closely related to barley and wheat. Rye is a cereal and should not be confused with ryegrass used for lawns, pasture, and hay for livestock. The slender seed spikes carry stiff awns with the grains growing in pairs. Unlike other cereals, the flowers open to pollenate. This allows much natural cross-fertilisation to take place, making it very difficult to keep strains of rye pure.

It grows better in poorer soils than any other cereal and is particularly tolerant of acid soils. It thrives in colder climates although it is not quite as hardy as barley. Existing

Fig 14: Rye

in an autumn variety only, it is planted at 80 to 110 pounds per acre in rows 6" to 7" apart and harvested when dry and yellowish grey in colour. Average British yields are around 15 cwt per acre but yields 2 to 3 times this are often obtained in some European countries. The low yields obtained have led some to refer to rye as "glorified grass". Rye has as great a food value as wheat but contains less gluten. This limits the action of yeasts which makes rye breads denser and also darker. It can also be eaten whole, either as boiled rye berries, or by being rolled, rather like rolled oats.

It is widely used for livestock feed in the form of rye hay. Medium sized particles produced as a by-product of milling called middlings, and the removed husks themselves known as bran, are also used as fodder.

Rye is widely used for the production of alcoholic beverages such as rye beer, as well as forming the basis for distillation as rye whiskey, Holland gin, and some vodkas. As with other cereals, these uses attract the highest prices.

The straw is long, smooth, and easy to bend so is used for making straw hats, mats, mattress stuffing, packing material, paper, and even corn dollies. It decays less than other straws and so is very popular for thatching.

Buckwheat

Family: *Polygonaceae* Genus: *Fagopyrum*
Varieties: *Fagopyrum Esculentum* and *Fagopyrum Tataricum*

Not a true cereal but rather a pseudocereal, buckwheat is cultivated primarily in America, Asia, and Eastern Europe. It grows about 3' tall with an erect central stem. It carries triangular or heart-shaped leaves, and white, pink, red, or greenish flowers. Honey from buckwheat is particularly dark and strong-tasting. The seeds produced are from 1/4" to 1/8" long.

Buckwheat grows best in a cool, moist climate. Growth of the crop is very rapid with harvesting possible after as little as 12 weeks.

The seeds are hulled to produce kernels known as groats. These are used for breakfast cereals and for making soups. In eastern Europe, coarsely crushed buckwheat is cooked to produce a mush known as kasha. It can also be milled into a flour which is used for making the traditional crepes of Brittany, as well as similar pancakes in America. When mixed with wheat flour, it is even used for making noodles in Asia.

Fig 15: Buckwheat

Triticale

Family: *Poaceae* Genus: *Triticosecale*

Triticale is a relatively recent hybrid of wheat and rye that contains more usable protein than either. The plant stands from 18" to 3' 6" high and carries 6 to 10 long narrow leaves. The head carries many spikelets each carrying 3 to 5 grain kernels. Cross-breeding began in 1876 but problems of sterility in the early strains were not finally cured until 1937.

Fig 16: Triticale

Dredge Corn

Dredge corn is a mixture of cereals specifically grown together for use as animal feed. It is used as an alternative to growing the individual items of the mixture separately and has the two following advantages. Firstly it eliminates the problem of storing materials separately and then having to mix them. Secondly, and perhaps more interestingly, it allows tall spindly cereals to be supported by shorter stouter ones. A popular choice would be a mix of 60% Barley with 40% oats.

Beans

Kingdom: *Plantae* Division: *Magnoliophyta* Class: *Magnoliopsida*
Order: *Fabales* Family: *Fabaceae* Tribe: *Viciae*
Genus: *Vicia* Species: ***Vicia faba***

Broad beans also variously known as butter beans, English beans, Windsor beans, horse beans, and fava beans. These are native to north Africa and south-west Asia, but are cultivated extensively elsewhere. They have a long tradition of cultivation in Old World agriculture, being among the most ancient plants in cultivation and also among the easiest to grow. It is believed that along with lentils, peas, and chickpeas, they became part of the eastern Mediterranean diet before 6000 BC. They are still often grown as a cover crop to prevent erosion because they can over-winter and because as a legume, they are able to fix nitrogen in the soil.

The bean plant is a rigid, erect plant 18" to 6' tall, with stout stems of square cross-section. The leaves are 4" to 10" long, pinnate with 2 to 7 leaflets, and of a distinctive grey-green colour. Unlike most other Vetches, the leaves do not have tendrils for climbing over other vegetation. The flowers are ½" to 1" long, with five petals. The main petal is all-white, the wing petals are white with a black spot.

The fruit is a broad leathery pod, green maturing blackish-brown, with a densely downy surface. In wild species, the pods are 2" to 4" long and ½" diameter, but modern cultivars developed for food use have pods 6" to 10" long and ¾" to 1¼" thick. Each pod contains 3-8 seeds round to oval in shape. These grow from ¼" to ¾" diameter in the wild plant. Some commercial varieties are even larger; ¾" to 1" long, ½" broad and ¼" to ½" thick.

Harvesting of a main crop sown in early spring takes place from mid to late summer. Horse beans left to mature fully are usually harvested in the late autumn.

Clover

Family: *Leguminous - Fabaceae*		Genus: *Trifolium*
Species:	White = *Repens*	Red = *Pratense*
	Strawberry = *Fragiferum*	Crimson = *Incarnatum*

Clovers are members of the pea family or legumes of which there over 300 varieties. They grow anything from 6" to 36" high, carry leaflets in groups of 3 to 6, and produce clusters of tiny flowers that can be in groups of anything between 5 and 200 coloured white, yellow, or any number of shades of red. Clover is chiefly used as a pasture crop but can be used for making hay or silage. It is also a valuable rotation crop which can increase the nitrogen content of the soil. Many, but not all, species are annuals and require planting every year. Hence there is a need for the commercial production of seed and this gives another task for the threshing machine.

Fig 17: Clover

Smut

Division: *Ustilaginomycetes* Order: *Ustinaginales*
Families: *Ustilaginaceae and TilletiaTeae*

All cereals can be subject to a group of parasitic conimycetous fungi collectively known as smut because of the resemblance of their black spores to soot. The fungus overwinters as spores in the soil or in manure and can survive like this for 2 or 3 years. The spores may be carried by the wind for long distances, and hail can provide open wounds which can greatly increase infection rates. Being of softer tissue, younger plants are particularly susceptible. More prevalent in damp seasons, it starts as black spots on the leaves and affects the performance of the plant. The spores can be blown to adjoining plants where the infection is repeated. If left untreated, the fugus can even take the place of the farinaceous material within the husk. Different varieties are adapted to different types of host plant, the most common being *uredo segetum* and *uredo foetida*. In recent times, it has been successfully treated with fungicides, as has the similar pest, Rinkasporum, which occurs more commonly in dry seasons. Part of the finishing process in the mechanised threshing process is specifically designed to remove smut balls. Not surprisingly, this apparatus is called a smutter.

Fig 18: Common Smut on Maize

Straw and Thatching

Fig 19: Bundles of Combed Reed Thatching Straw

As the threshing machine was improved, British manufacturers sought to produce not only the maximum amount of quality grain, but also the maximum amount of quality straw. This could be cut into chaff for use as animal feed, especially for horses. It could be used as an animal bedding material in which case a better quality sample was more saleable. But the most demanding market was for use in thatching where only unbroken stems were usable. Traditionally, Britain has used three thatching materials. The best of these is water reed grown in East Anglia, and with a life of up to 70 years. This has been harvested commercially as a crop and one man can cut between 50 and 100 sheaves of reed in a day. Northern England & Scotland made use of heather, the poorest of the three, and Southern England used straw. This only gave a 20 to 40 year life so the total amount of material required would be much larger than that used for the same number of roofs thatched in reed. Local planning authorities often insist on straw being replaced with straw rather than the longer-lasting reed. It has been said that if it is dressed correctly, after a few months, it is impossible to tell the difference.

Improved transportation firstly in the form of canals, and later railways resulted in a major increase in Welsh slate production from 1820, matched soon after by a similar increase in clay tile production. These longer-lasting roof coverings sounded the death-knell for thatching and 1,000,000 houses with thatched roofs in 1800 had dropped to a mere 35,000 by 1960. The reasons for retaining thatch on these houses would be mainly aesthetic rather than economic. When re-roofing was required, the better quality and longer life of reed would make this the material of choice, making the amount of straw required even less. This change in choice of roofing material combined with a reduction in the number of horses made straw virtually worthless.

The combine harvester is able to process larger quantities of grain if straw throughput can be reduced. This is achieved by raising the cutter deck as high as possible to leave a long stubble. Parallel changes in cereal varieties have also reduced the overall length of the straw. This means that such straw is not suitable for use in thatching being broken and just too short in the first place. In the Home Counties, straw for thatching is wetted and then drawn into small bundles to allow straightening called yealms. These are 4" thick and 14" to 18" long for use in a method known as long straw thatching. In the south-west, larger bundles known as "nitches" are kept dry and put through a machine known as a reed comber. Although the material is still straw, this is known as combed reed thatching.

Today, thatching a roof in straw requires a complete return to the past. Different varieties of cereal have to be specially grown, to be then cut using a binder prior to being processed using either a threshing machine or a reed comber. Favoured older varieties include Little Josh, Maris Wigeon, and Huntsman. The National Trust has been carrying out a research into older straw varieties by taking samples of grain for straw used on very old roofs with a view to cultivating it once again in a sort of botanical "Jurassic Park" programme. Triticale (Triticum Turgidum) is a relatively modern hybrid whose rye origins give it longer stiffer stems. A more modern hybrid is N59. All these are winter sown and are best after dry summers being strong, pliable, and long. Durum pasta wheat is very tough but brittle when solid. It is much more pliable if hollow. Rye straw is used sometimes but generally only for ridging. The amount of nitrate fertilizer used on these cereals has to be limited to retain stem stiffness. As an aside, the Norfolk reed now used on almost 50% of British thatched buildings is now cultivated specifically, with 90% imported from Turkey and various parts of Eastern Europe including Romania, Hungary, and the Ukraine.

The Anglo-Saxon language and imperial measurements come into play again here. When describing thatching, 8 sheaves make 1 stook, 2 stooks make 1 stock. An imperial bundle is 24" in circumference as measured at the butt end. Alternatively, they can be described in terms of how many bundles can be encircled by a 6' length of a rope. The would be typically referred to as "7 to the fathom". An East Anglian "Nitch" is a bundle of reed weighing 28 lbs.

Fig 20: Combed reed thatching

Chapter Three

SCHEMATIC DESIGN

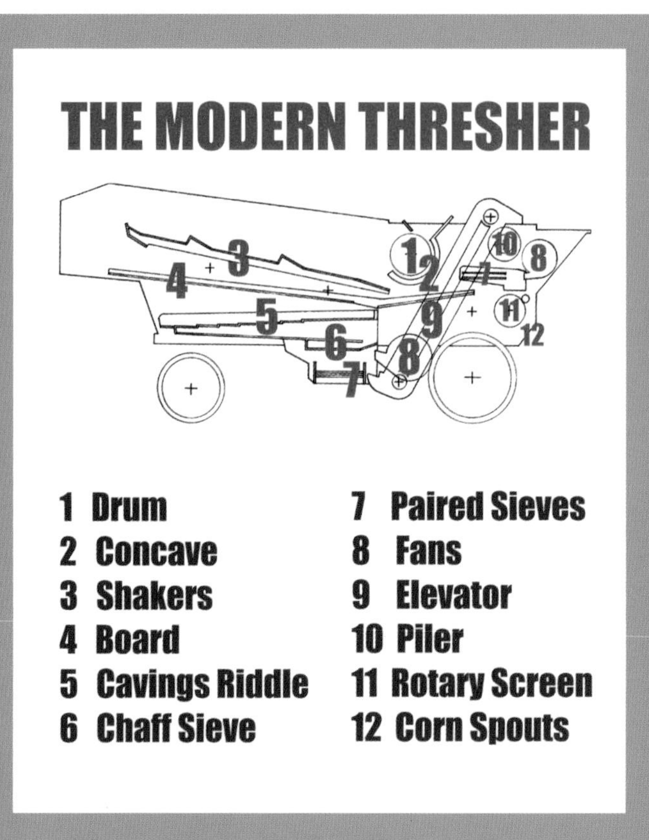

Fig 21: The Modern Thresher - Principal Parts

The Modern Threshing Machine

If the measure of the effectiveness of a threshing method is measured by the proportion of the crop salvaged, the separation of other material, and the quality of the finished article, then the modern threshing machine reigns supreme. Unfortunately, this is at the expense of a level of manpower which, in spite of the above merits, makes the system completely uneconomic and which has finally consigned the thresher to history.

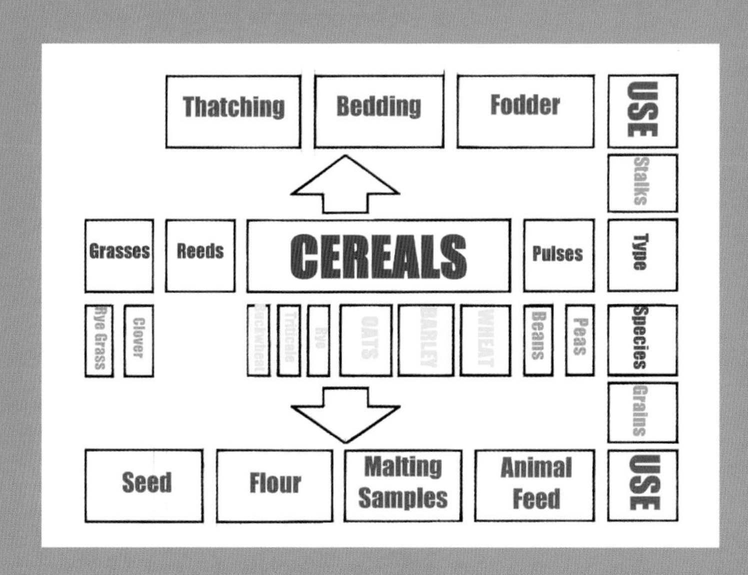

Fig 22: Diagram - Separation & Uses

The object of the exercise is to divide the crop, plus lots of other unwanted material mixed with it, into its different constituent parts and deliver each completely separately. An 1881 article, extolling the virtues of the thresher, states that "In foreign countries, the quantity of weeds collected with the crop is sometimes more than the crop itself". The action of the threshing machine can be seen as involving three distinct operations. The first is the detachment of certain parts of the crop by mechanical means. The second is agitation of the mixture to discourage the different parts from sticking together. The third is the sorting of all of the constituent parts so that they can be collected individually. Writing in the early 1880's William Worby Beaumont stated that "This is thus a somewhat complicated piece of combined mechanism, which is purely the result of experiment, unaided by any direct application of theory". His thoughts were echoed by a number of early 20th century writers. One described the machine in the following terms: "Apparently simple, it is essentially scientific. Its dynamics abound with interesting surprises. Its success depends upon the satisfactory solution of numerous problems of machines", and, "There is perhaps no single machine designed to carry out any series of operations or processes, in which so many conditions and circumstances are involved, and have to be fully considered, and provided for, than the finishing threshing machine".

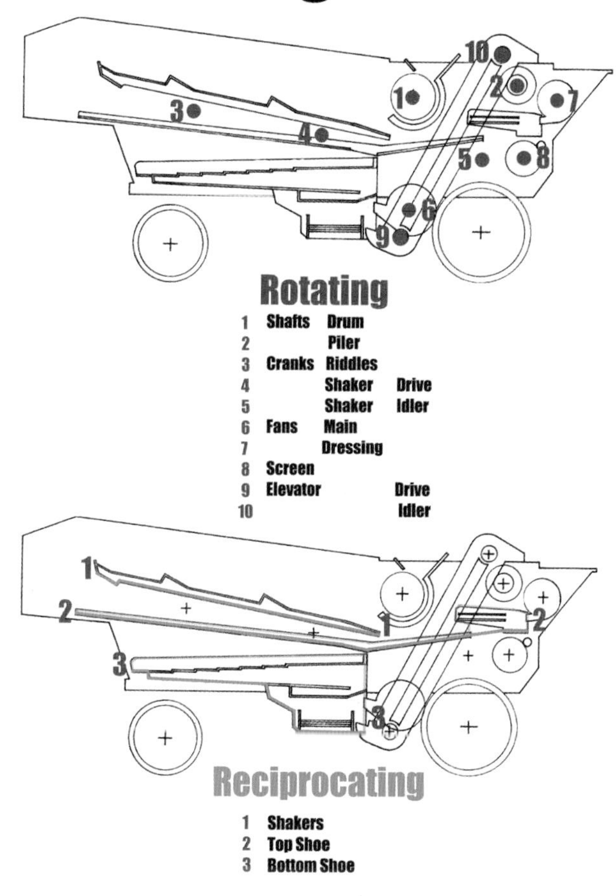

Fig 23: Moving parts

The first of the thresher's operations is achieved by a combination of beating and rubbing. The second is by using various forms of reciprocating motion. The third is achieved in two different ways. Primarily by sifting according to size which involves passing the various parts of the crop over a number of grids and gratings, riddles and sieves. This is supplemented by the use of blasts of air to separate the lighter material by weight rather than size, the process known as winnowing.

PART ONE

In the simplest terms, the thresher is a large rectangular box on wheels which has the crop, either direct from the field, or from the rick, fed in at the top. From one end emerges grain, and from the other end emerges straw. Because the modern threshing machine was originally designed with steam as the motive power in mind, discharge of the more bulky, and more importantly inflammable straw will be away from the engine. This means that the corn will be collected from in front of the engine's smokebox. The engine thus faces the corn end of the machine, which is usually built to be towed from the straw end. Thus the terms offside and nearside for the machine will be the same as those applied to the engine.

Many early machines discharged the corn from the side of the machine requiring a 90 degree change of direction, not only for the grain, but also for the drive to a group of components. This seems an unnecessary complication when compared to the much more straightforward design with "all parallel shafts", but interestingly early combine harvesters also adopted this same side-delivery design

Fig 24: 1875 Humphries Machine

Fig 25: Bevel drive to Rotary Screen

But the modern thresher is very much more sophisticated than its simple box design suggests. Apart from the straw and varying qualities of grain, a whole variety of other materials are released separately through a number of other outlets located at various points along the underside of the machine, to be recycled, or disposed of.

Drum

Fig 26: Drum

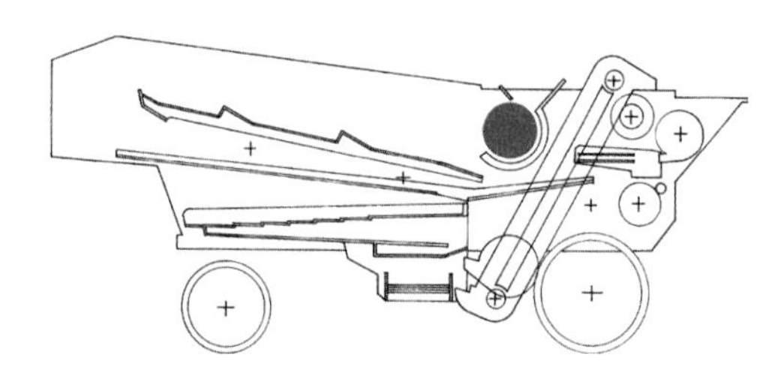

The threshing drum carries a set of beater bars, usually 8 but sometimes 6 and rotates at between 1000 and 1100 rpm. With a diameter of nearly 2', this results in a beater bar speed of between 60 and 70 mph. Because of this, on all but the earliest machines, the opening to the drum is fitted with a safety guard which is available in a whole range of designs. This is not for the protection of the person actually feeding the machine, but for the benefit of the other operatives standing on the threshing platform. Alternatively, a self-feeder mechanism may be fitted which usually takes the form of a short elevator. In order for the machine to run efficiently, a constant speed needs to be maintained. This is difficult because however skilled the person feeding, the flow of sheaves is bound to vary. A partial solution to this problem is to make the drum very heavy so that it acts as a flywheel. Also there is a governor on the power source which attempts to keep the speed constant despite the fluctuating load.

The different power sources are equipped with different sized flywheels or pulleys to maintain the belt speed of 2100 feet per minute required to turn the drive of the 8" diameter pulley on the thresher's mainshaft. This means that the slower the rotational speed of the motive power, the larger the pulley or flywheel required to maintain the magic 1000 to 1100 rpm. It is even possible to deduce the optimum operating speed of the motive power source from the size of its drive pulley.

The crop is subjected not only to simple beating. The gradual narrowing of the gap between drum and concave also results in a measure of rubbing action. As well as separating the different parts of the crop, and not to mention everything else fed into the machine with it, the straw is broken to a greater or lesser extent.

Rather like the Eskimo who is supposed to have so many Inuit words for snow, so the farmer has a whole range of Anglo-Saxon terms for straw. The unbroken long stalks are still referred to as *straw*. Larger broken pieces of straw are known as *chibbles*. Smaller broken pieces of straw or leaf are known as *cavings*. Even smaller broken pieces of straw are known as *chaff*.

Once the mechanical separation has taken place, it is necessary to sift this mixture of materials. Essentially this means passing the mixture over a series of slots and holes that become progressively smaller.

Concave

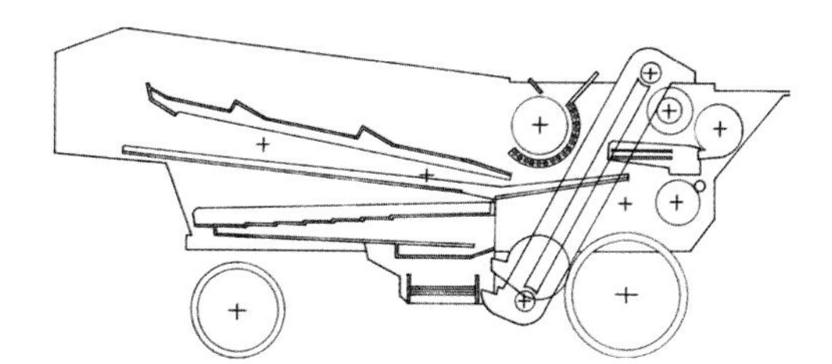

Fig 27: Concave

The first of these are the rectangular slots in the concave formed by the metal cross members and the wires threaded through them. These are generally in the order of ¼" wide and 1½" long. It is intended that ALL the straw is retained by the concave whilst allowing as much of the grain as possible to pass through. This is helped by the first of a number of hanging boards at the beginning of the shakers that direct flying grains back towards the concave.

Shakers

Fig 28: Grain Board

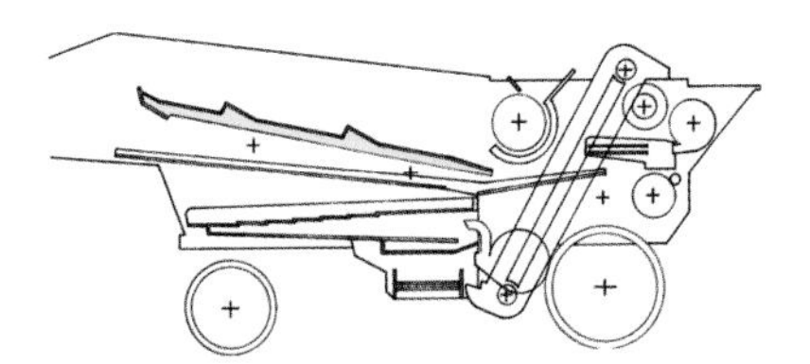

A second set of rectangular slots are to be found in the shakers. These are formed by the individual shaker frame side members and cross spars. These slots are typically 11" wide and can vary between 3/8" and 1" long. Just like the concave, these are intended to retain the clean unbroken straw whilst allowing all other material to pass through. At the end of the shakers, the straw passes out of the end of the machine. One or two sets of boards similar to those beside the drum are positioned along the shakers. These are held on check chains which allow them be adjusted so as to slow the passage of the straw, giving more time for trapped grains to fall to below. When adjusted correctly, these produce a thin even "mat" of straw.

Grain Board

Fig 29: Grain board

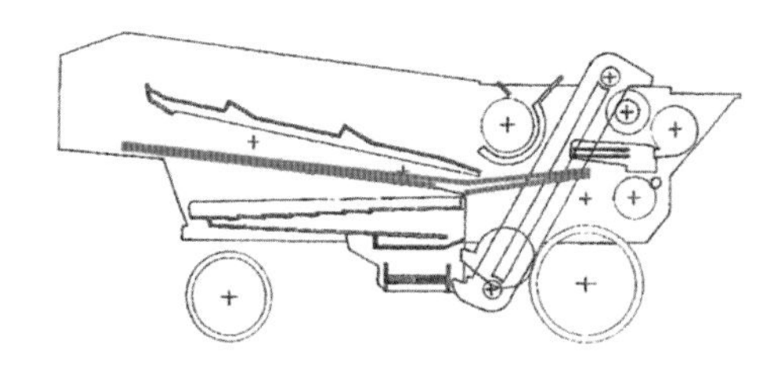

All the material that falls through the slots in both concave and shakers lands on the reciprocating grain board, where it is joined by other material from the dressing shoe for a second pass through the machine, of which more later.

Cavings Riddle

Fig 30: Cavings Riddle

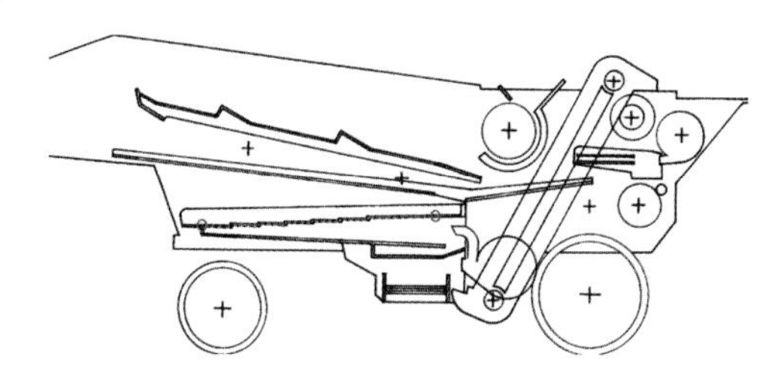

The V-shaped grain board conducts material to the main shoe, the third point of sifting. This carried out by a series of sieves, the number of which can vary. The first is always the cavings riddle. This is a continuous row of screen panels usually arranged in a gentle stepped pattern. It is intended that the riddle retains all the cavings so that they pass out of the end of the machine below the straw, whilst allowing all other material to pass through. This set of screens is not interchangeable. To keep the Cavings separate from the Straw, a loose board is often placed below the ends of the Shakers.

At this point, the only really valuable material left is the grain and from here on, the principal requirement is to deliver this in the best possible condition. This is achieved by gradually removing the unwanted material mixed with it by passing it over a series of screens with holes of decreasing size.

Main Shoe

Fig 31: Chaff Sieve

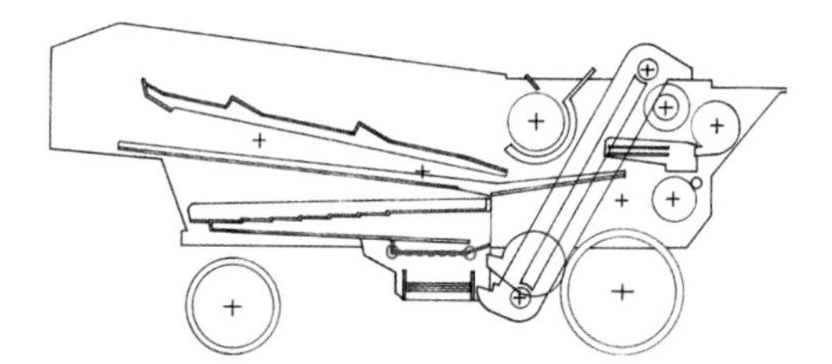

The next point of sifting is a medium-sized screen with medium-sized holes, slightly angled, or with a low bar alongside. This is the principal screen and like the cavings riddle, is not generally interchangeable.

In the bottom of the main shoe is located either a one or a pair of primary sieves. If there are two then these are located one above the other, with the holes in the upper sieve being slightly larger than those in the lower. Unlike their predecessors, they are specifically designed to be easily changed to suit the different crops being processed.

Sieves

Fig 32: Sieves

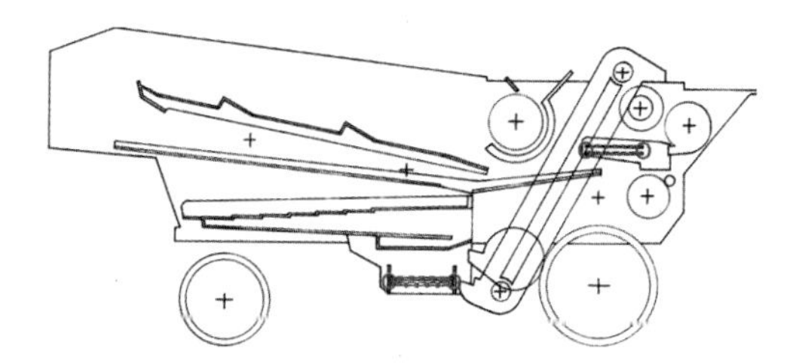

These sieves are intended to remove all the remaining material LARGER than the grain. Thus the sieve or sieves chosen will have holes that are only slightly larger than the relevant grain size. The material retained by these sieves is known collectively as chobs. These include thistle heads, broken heads of grain, plus larger pieces of debris and other unwanted material. But it also contains some unthreshed grain. All this material is discharged separately to be collected in a basket and it can be fed into the drum for a second time. This is another example of how efficient the threshing machine can be in terms of the proportion of grain available that can be reclaimed. The final sieve in the very bottom of the shoe is the dust sieve. This is intended to release all the remaining material *smaller* than the grain and has tiny holes. Indeed sometimes this screen is replaced with a blank plate. The material that falls through the dust sieve onto the ground below the machine can be collected in baskets. This can be fed to poultry or taken away for burning. The emptying of these baskets is often described as the least pleasant of jobs associated with the threshing machine.

Elevator

Fig 33: Elevator

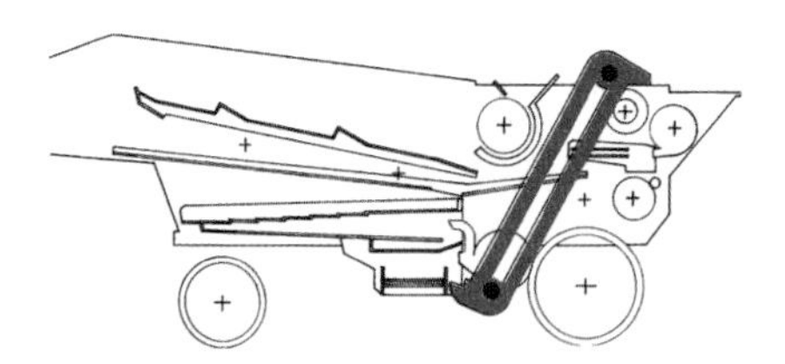

At this point, the primary action of the machine is complete. Up to now, the movement of the grain has been by gravity, albeit assisted by reciprocating motion. Now the elevator, or in a few cases possibly a blower, takes the material to the top of the machine for a second time.

Spirit Level

PART TWO

From the top of the top of the elevator, the path of the grain is controlled by a series of shutters that allow a choice of paths. A short auger takes the grain to the first of these shutters. Opening this first shutter allows the grain to bypass the entire dressing mechanism. In this position, all the grain drops directly to the two hopper-shaped compartments above the two best corn spouts from which it can be bagged. If it is intended to present the individual kernels of grain clean, polished, and free of all other material, then these require further mechanical treatment. There are a number of materials to be removed. These include the small attached fibres known collectively as *awns* which are typified by the *"beards"* on barley, and the husks or *"whitecoats"* found on wheat. Also there are the soot-like spores of a number of species of fungus collectively known as *smut.*

Piler

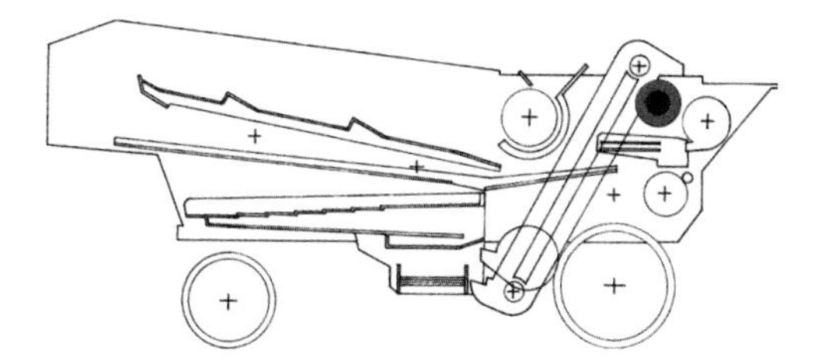

Fig 34: Piler

The feed auger is mounted on the end of a shaft that also carries a number of blades rotating in a cylinder. This mechanism is collectively known as the piler. If the drum is designed to beat and rub the crop, then the piler is designed to rub and beat the grain. A system of shutters and flaps controls the passage of the grain through the piler. The flaps allow the grain to fall from the instrument at a choice of points along its length. The shutters slow the flow of the grain through the instrument. This works on the principle that greater cleaning is obtained by longer exposure to the various blades.

All the grain that passes through any part of the piler mechanism is then directed to another single screen or pair of screens in the dressing shoe similar in design to the sieves of the main shoe. These are also designed to be interchangeable according to crop type. They are of a smaller size overall than those in the main shoe. The hole size will usually be the same as in the main shoe, or it may be slightly smaller. The intention behind these two points of sifting is to remove the material detached by the piler. They allow the grain to pass through whilst directing all larger material back to the grain board from where, as with the material from the chob sieve, it is passed through the machine for sieving a second time. Finally below these screens lies the last point of sifting, a second dust screen. This retains the grain, discharging the smaller material onto the ground below the machine, where once it can be collected in baskets. As in the main shoe, this too is sometimes replaced by a blank plate.

Rotary Screen

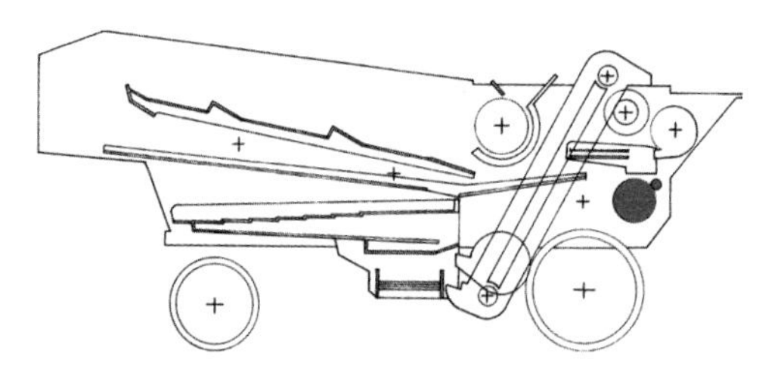

Fig 35: Rotary Screen

A board below the dressing screens directs the finished grain to a shaft carrying a second short auger leading to the rotary screen. This consists of a long cylinder of wire fitted with a light flexible plate, formed into a spiral shape, which moves the grain along its length. The screen is so arranged so that along its length, the gaps between the wires become increasingly large. Below the screen, a number of hopper-shaped compartments lead to the corn spouts to which the sacks are attached. The first of these, under the narrowest gaps between the wires, collects dust. The next one, or occasionally two, collects tailings or second corn. The last is double-sized so as to feed two corn spouts and these are for best corn. It is these spouts that will collect the corn if the dressing mechanism is bypassed. The two main corn spouts are opened by a pair of vertical sliding shutters linked by an inverted T-shaped bar with a handle which closes one spout as the other is opened. This allows the sacks to be easily changed whilst maintaining a constant flow of grain. Similar shutters are fitted to the corn spouts for the dust and tail corn, but as these fill very much more slowly, the shutters can be left closed for a short time while these sacks are changed.

Winnowing

In addition to the sifting systems described above that operate according to the *size* of material, the machine also uses the traditional winnowing principle to remove less dense material using blasts of air. The main fan produces the first blast that passes through a short duct directed upwards towards the underside of the principal sieve.

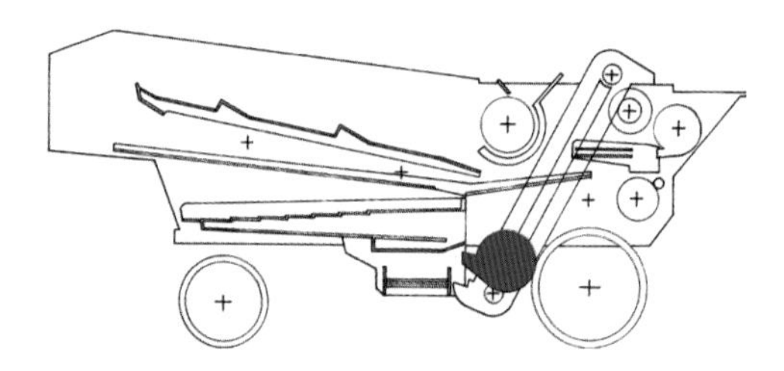

Fig 36: Main Fan

The smaller dressing fan produces the second blast that is ducted similarly between the two secondary or dressing sieves and to be directed upwards towards the underside of the upper sieve.

It is not uncommon for some machines to incorporate a third blast directed upwards towards the underside of the cavings riddle. Its principal purpose is not so much to sort material as to prevent blockages occurring. A few machines even feature a fourth blast directed at the end of the grain board just beyond the dressing shoe. As with the third blast, this is not to sort material, but to prevent blockages.

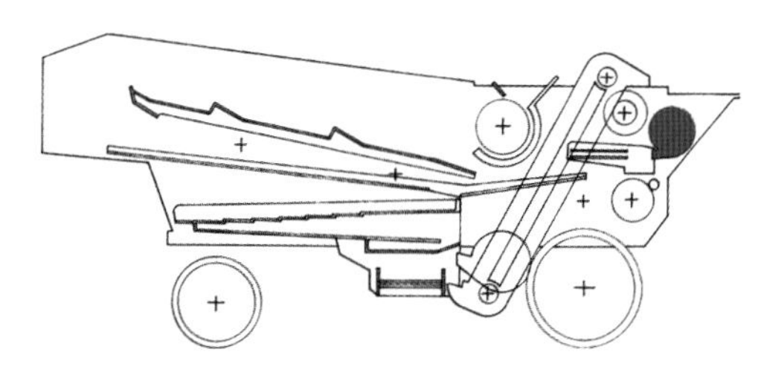

Fig 37: Dressing Fan

The earliest threshing machines fall into two distinct groups; fixed and portable. Andrew Meikle was right when he said that effective threshing was dependant on striking the crop sufficiently smartly, which means a high beater speed. Until the arrival of steam as motive power, all machines were handicapped by the slow rotational speed of the only effective power sources available, water wheels and horse-worked capstans. The fixed machines in barns were able to produce a moderate beater speed by the simple expedient of increasing the diameter of their slow-rotating drums. These machines also featured a large sieve below the threshing drum extended below other even larger drums that tossed the straw towards the outlet allowing some of the trapped grain to fall. This layout made them reasonably efficient for their time, but at the expense of being very large and cumbersome.

The earliest portable machines by comparison, were much smaller and were quite literally threshers. A small drum and concave detached the ears but all the other operations of winnowing and dressing were performed separately.... and manually. The efficiency of these machines greatly improved when the high rotational speed of the new portable steam engine enabled beater speeds to reach, and soon exceed, those of their fixed rivals. As these beater speeds increased, so it was discovered that it was no longer necessary to hold the crop between a pair of feed rollers in order to achieve sufficient impact from the beater bars. A major step forward in thresher design came with the introduction of the open concave.

The next innovation may be seen as the biggest single step forward in thresher design; the "bolting" thresher. The feed via a pair of rollers had meant that the crop had to be fed end-wise, usually "head-first". The action of the drum treated the straw very roughly, delivering it in such a broken state that it made it unsuitable for thatching. This meant that all straw for thatching at this period had to be the result of threshing by flail. The breaking of the straw by machine was increased by the action of the feed rollers. Some of these were plain cylinders but the preferred design was of a corrugated pattern. These dealt better with wet crops but at the expense of yet more damage to the straw. By the simple expedient of making the drum sufficiently wide, it was possible to feed the crop crossways in sheaves or "bolts" hence the name. By passing the straw in this way, damage was minimised and for the first time, it was possible to produce a straw sample by machine that was suitable for use in thatching. These unbroken samples produced also made them much more presentable for resale, and this was seen as a selling point to farmers who "served a town's need for straw".

This high quality of straw is matched by the grain samples produced. At one extreme, the mixture of cereals known as dredge corn can be threshed to give an acceptable animal feed without the use of the dressing mechanism. Barley can be dressed in such a way that short awns are left to speed up fermentation when used for brewing. At the other extreme, a fine polished sample can be produced that will satisfy the most demanding of millers. As well as the straw and grain, the chaff was discharged separately so that it could be easily collected used as animal feed. Not only that but even the small seeds and the like were discharged separately and these could be used as poultry feed, or disposed of to prevent the growth of weeds.

Although the basic portable thresher started off much smaller than its fixed rival, all the subsequent changes that started with the bolting principle made it larger and larger. It was found that, not unsurprisingly, the longer the shakers, the more opportunity there was for grain to be dislodged from the straw. This can be demonstrated graphically by comparing Garretts 1852 machine with the Ransomes 1926 design. If the circumference of the drum is taken as the reference figure of 100%, then the total maximum distance travelled by grain in each of the machines is 2.8 times the drum circumference for the Garrett, and 6.2 times for the Ransomes. Also the path of the grain in the Garret is 52% on reciprocating surfaces; on the Ransomes this is 93%; indeed the only fixed surface is the Concave.

Fig 38: Comparison - 1852 Garrett v 1920 Ransomes

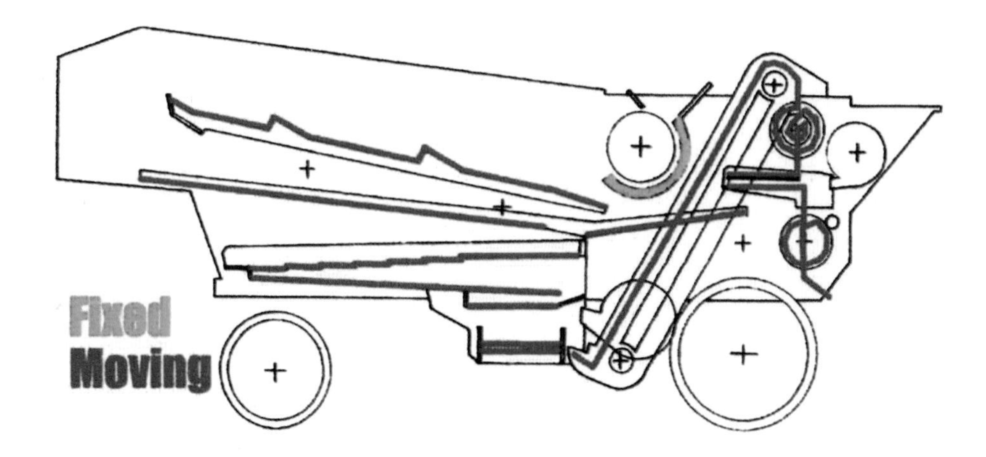

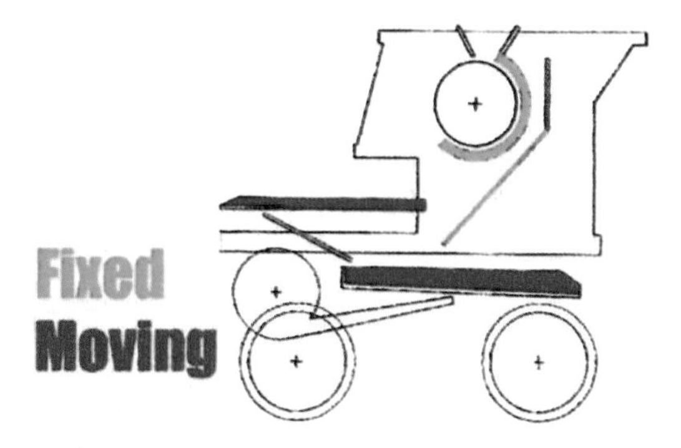

GRAIN PATH - RELATIVE TRAVEL DISTANCES - %'s		
Component	**Garrett 1852**	**Ransomes 1926**
Drum	100	100
Concave	37	30
Board	12	*
Shakers	36 R	100 R
Board	22	86 R
Screen	53 R	*
Cavings Riddle	*	67 R
Bottom Shoe	*	65 R
Chaff Sieve	*	14 R
Chob Sieve	*	30 R
Total	**160**	**392**
Reciprocating (R)	**56%**	**92%**

Table 2: Comparison - 1852 Garrett v 1920 Ransomes

Efficiency was also improved by fitting a greater number of screens in a variety of sizes, all made as large as practicable. The size of fans also increased. This is based on the simple hypothesis that the force of the blast can always be reduced. By simply closing the fan shutter doors, it is possible to achieve a reduction in the order of a factor of 3. To improve quality of output still further, a dressing mechanism was added which was not unlike a whole second machine, albeit on a smaller scale. Finally a grading system was added in the form of the rotary screen.

The machine managed to remain portable… but had grown substantially.

Chapter Four

CONSTRUCTION

Drum-Speed Plate (Garvie)

The typical modern threshing machine may be thought of a rectangular box on wheels. The majority of the components inside extend the full width of the interior of the box. The machines were made in a range of sizes, this being related to the width of the drum. All sizes were based on a 6" increment, from 30" to 66". The most common British size was the 54". This gave the machine an overall width of 8' 0" which was seen as being the widest machine that could fit through a normal English gateway and be towed down a narrow lane. The narrow lanes of the West Country and Ireland were generally thought of as being the limiting factor. Most larger machines would usually be for export, often to Eastern Europe. A number of smaller machines down to 36" were also popular and these were often referred to as "farmers' machines". Some of these were also exported, a major market being Argentina.

Fig 39: Typical Machine (Ransomes)

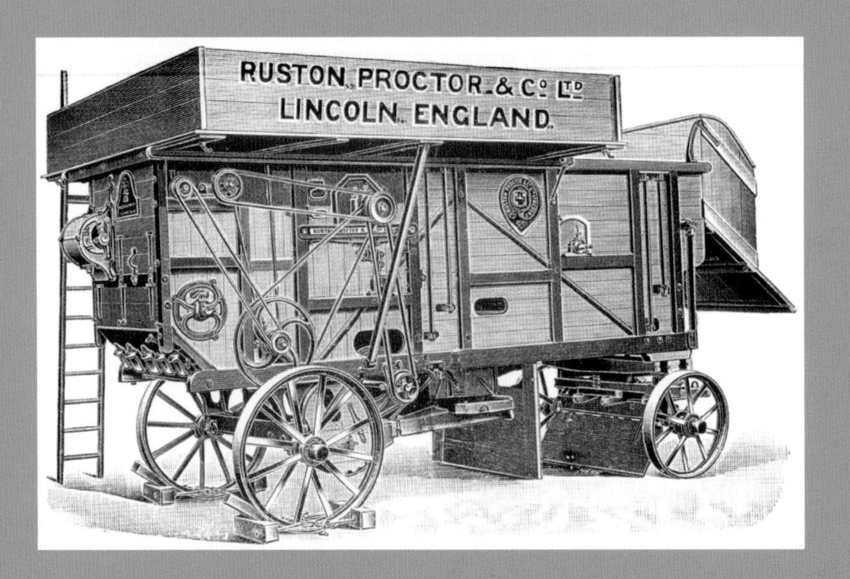

Drum diameters were either 20", 22" or 24", the most popular British size being 22". Although the size of machines is usually described in terms of drum width, because of the nature of the construction of the machine, this would apply equally to the interior width of the box and all the other full-width components inside. Some earlier machines mounted the components towards the end of the process at 90 degrees but all later machines have all the shafts parallel with the feed at the top, the straw output at one end, and the corn output at the other. A typical machine would be constructed as follows:

Box

The box is framed in seasoned timber beginning with bottom rails or sills of 6" x 3" section. Top rails were of 4" x 3" section and these and the sills would each be mortised to take the stub-tenons of the 4½" x 3" uprights. The 6" x 3¼" cross beams may be tenoned in the same way or bolted via cast-iron brackets.

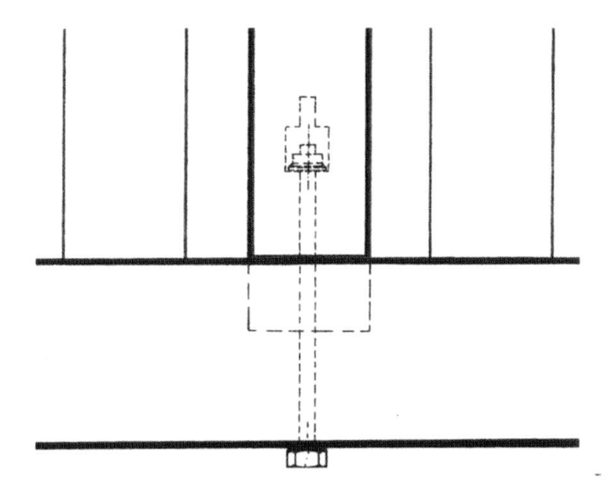

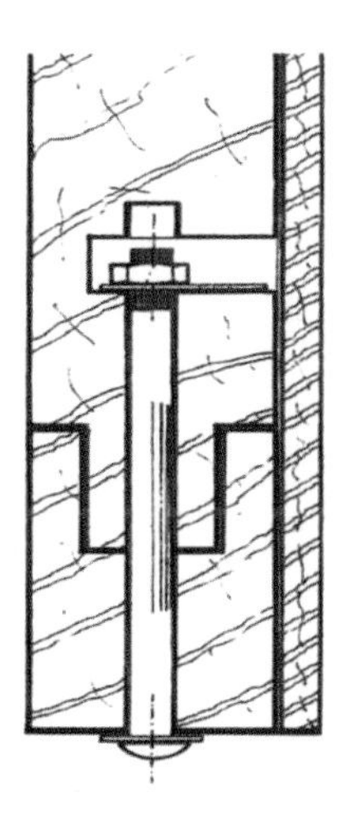

Fig 40: Bottom Sill Joint - Elevation

Fig 41: Bottom Sill Joint - Section

The framing that extends upwards above the end of the shakers is of smaller 3" x 3" section. The various frame members are coach-bolted together with concealed joints, and the uprights are suitably notched to receive the numerous cast-iron brackets that hold the bearings for the shafts. All coach bolts would be of Whitworth thread with large diameter washers. The nuts would all be hexagonal rather than the cheaper square pattern used more commonly at the time for joinery. The timbers of choice would be oak or ash and a reputable manufacturer, like Burrells for example, would be using stock that had been naturally air-dried in a shed for four years or even more. As a rule of thumb, the sawyers at Taskers reckoned to allow 12 months seasoning for every inch thickness of timber. Scottish manufacturers sometimes used home-grown fir. After the First World War, availability of seasoned timber fell sharply and many later machines were built using pitch pine. This had been a popular material for export machines being not only cheaper, but also lighter which would have reduced shipping costs. Some of the last machines were built with a steel frame, but there was a reluctance on the part of some manufacturers to change. Many early machines designed for export had been built with steel frames and had gained a reputation for twisting during shipment and arriving with a permanent set out of line.

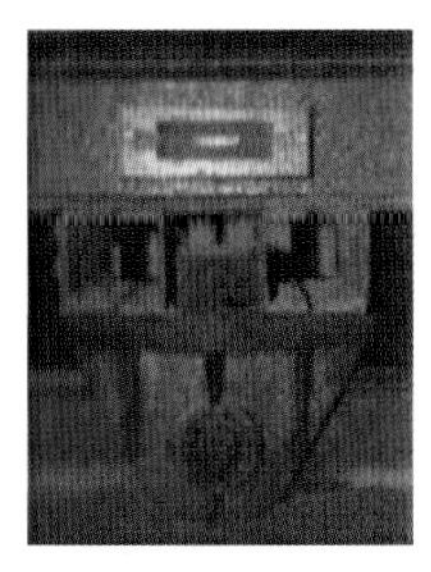

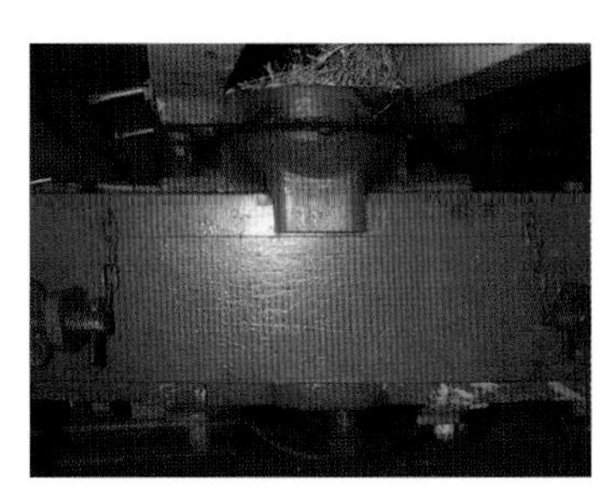

Fig 42: Forecarriage Pivot (Tullos)

Fig 43: Forecarriage Pivot (Garvie)

Fig 44: Shock Springs

The sides of the box would be clad in 4" x ¾" tongue-and-groove boarding often of redwood deal. Size of the box overall would be around 17' long and 5' high extending upwards another 18" or so over the end of the shakers. Sometimes this was extended lengthwise to cover longer shakers. This was referred to as a wind hood. If the machine was built by a traction engine manufacturer, then it would have built in the same shop that assembled living vans and wagons.

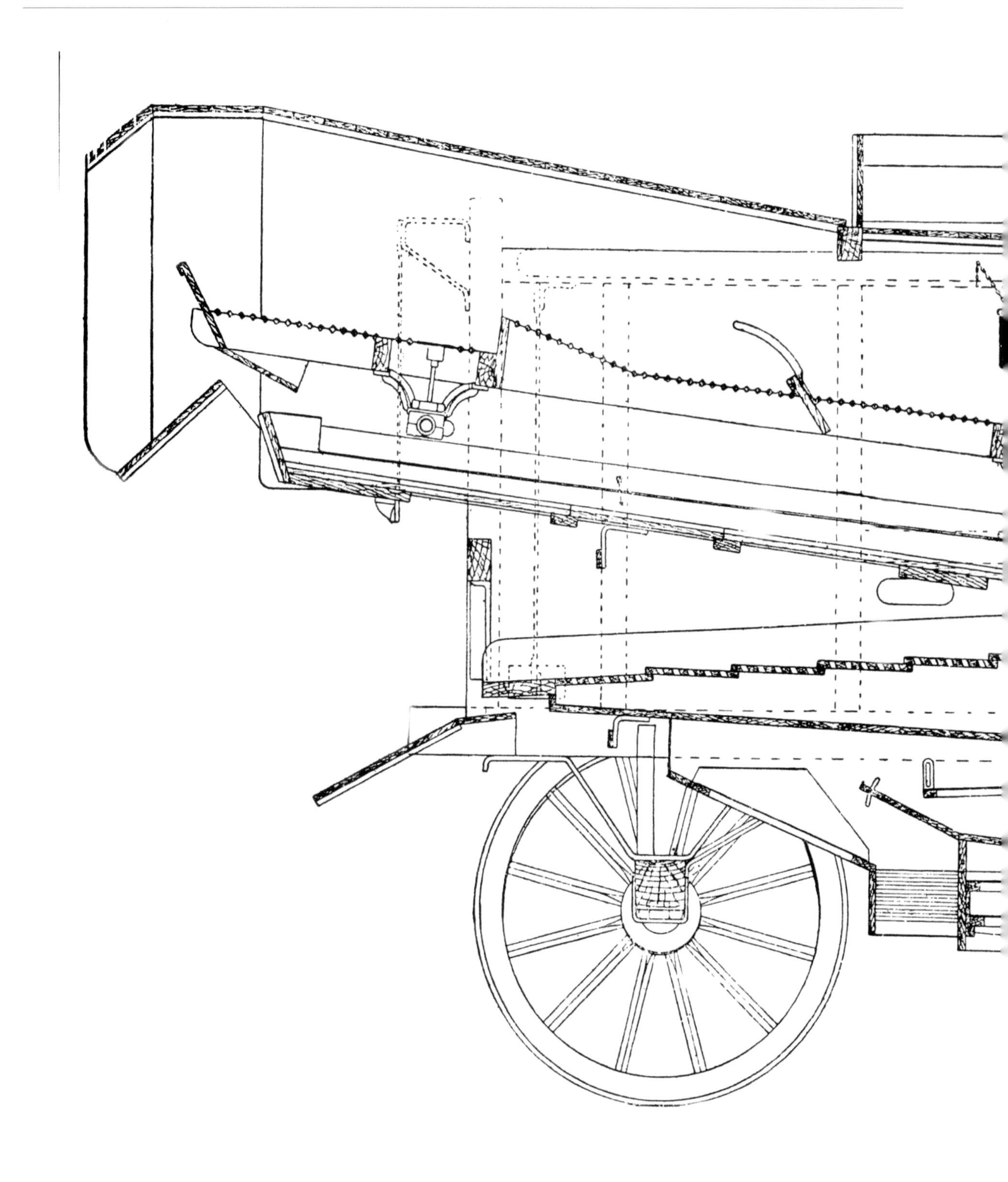

Section

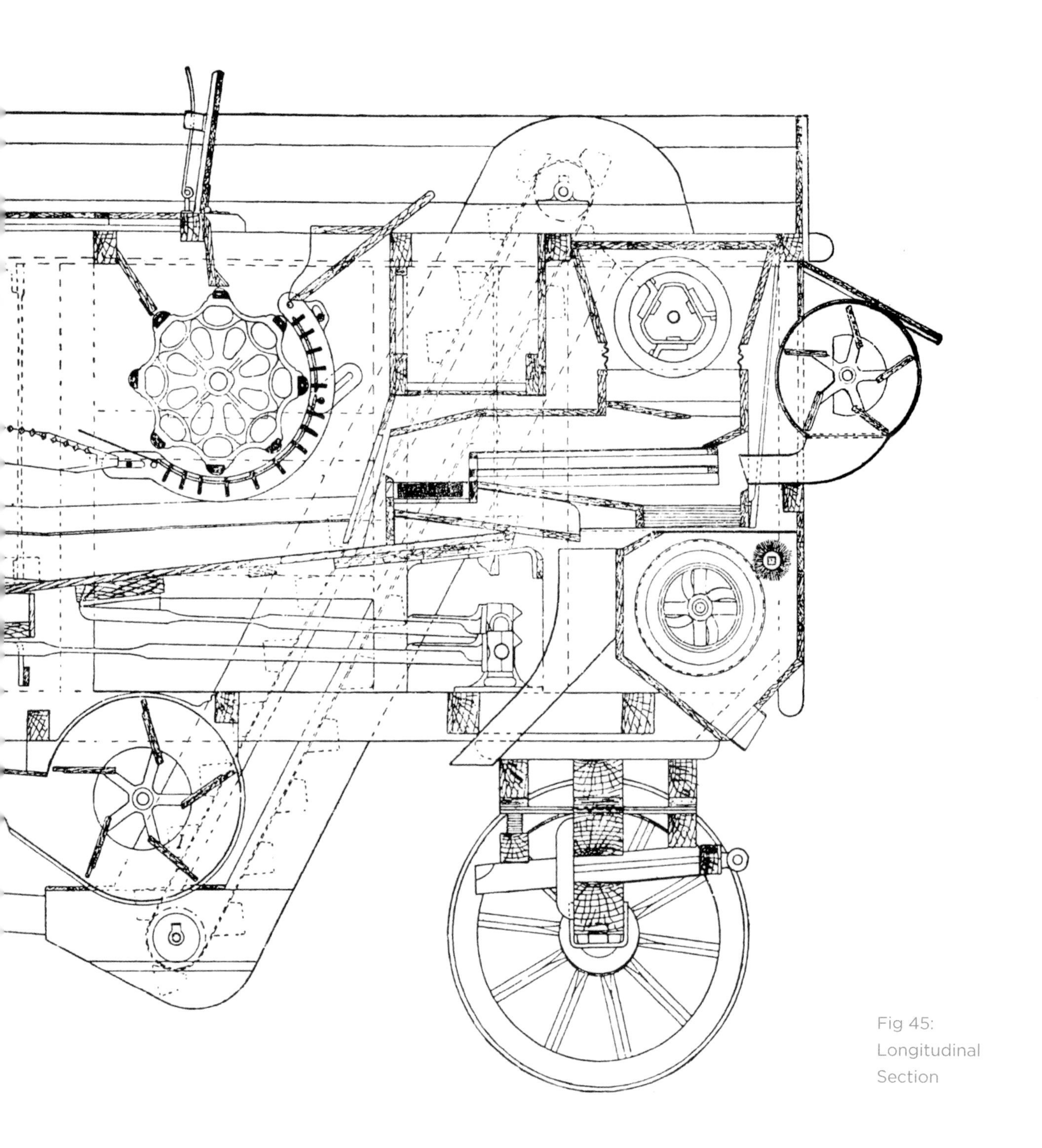

Fig 45: Longitudinal Section

Platform

The centre of the platform was basically level with two exceptions. Immediately in front of the drum was located the feeding box where the feeder stood. This extended the full width of the drum and was around 2' across and 2' deep. The front of any wind hood began to slope up some distance beyond the drum to give additional clearance above the shakers. The platform was also fitted with a number of full-width access panels. Each of these was hinged along its long edge and opened using a single ring handle that was rebated into the lid so as not to present a trip hazard. The first panel was located immediately behind the feeder and opened away from it to access the piler. The second panel lay beyond the drum feed and also opened away from the feeder, this time to access the far end of the drum. The other panel, or series of panels were above the shakers and hinged in the same direction as the drum access panel. When threshing, the machine was worked from an extending timber platform. A wing board was fitted to each side of the machine hanging down on long strap-pattern hinges. These boards were lifted up to the horizontal position and propped up on a 1" diameter iron rods around 5' long rather like broomsticks. Their lower ends sat in small cast iron brackets. Their top ends were sometimes attached to the boards via ring bolts, whereas some builders elected to keep the support rods loose and fitted the same pattern bottom brackets onto the underside of the platform.

Fig 46: Platform in extended position (Foster)

Fig 47: Platform bottom support bracket (Garvie)

A pair of vertical boards were slotted into the front and back ends of the extended threshing platform. Another pair of similar boards were then fitted to the sides. On the opposite side to the rick, the board would be fitted vertically. On the rick side, the board would be fitted horizontally, suspended on short chains from the end boards to extend the platform further. All the boarding for the platform assembly was from 6" x ½" tongue and groove timber. The vertical boards were the largest uninterrupted vertical surfaces on the machine. These gave the best opportunity for advertisement. They would be sign-written, usually with the name of the maker, but sometimes with that of the contractor.

Some later machines were fitted with conventional modern pattern internal-expanding drum brakes but earlier models used external-contracting brands operating on a drum bolted to each wheel on the fixed axle, with a cross shaft and a long handle passing through a slot. The slot was drilled with a number of holes so that a pin could be inserted to lock the mechanism. But this was intended only as a parking brake. The machine still required chocks or chain-blocks to hold it firmly in position when threshing.

The axles were of timber with steel stub axle ends. The fixed axle was usually mounted on simple bent flat steel straps, but some manufacturers elected to use cast iron brackets instead. The steering axle or forecarriage was arranged to rotate and pivot in the same manner as a conventional wagon. Unlike most manufacturers, Ransomes and Garvie elected to use a shallow ball-and-socket type arrangement. Whichever approach was adopted, the axle had to be able to swivel sideways but not forward and backwards. This was not always achieved as successfully as it might have been. Both axles were built very substantially to withstand the stresses caused by winching and the use of push poles but this could cause damage to the forecarriage pivot. When traction engines were the normal means of motive power, the towbar was often attached via a pair of springs to reduce the shocks to the forecarriage.

Mechanism

As previously mentioned, all the rotating parts run parallel across the machine and their bearings are located in iron brackets, cast in sand, and coach bolted into the frame members which are often suitably notched to receive them. The only machining to the shafts is for the bearing surfaces and for the keyways. As many bearings as possible were fitted externally to make, lubrication, adjustment, and replacement easier. Early machines used the same layout for the mainshaft bearings as used on all the other shafts. The additional loading was controlled by incorporating long flat metal straps bolted to the timber framing. Later, on all but the smallest machines, the pair of brackets to hold the drum bearings and the concave adjustment were large heavy castings usually carrying the maker's name. Castings would be produced using wooden patterns pressed into fine black sand with a binding material, often horse manure. The men filling the mould boxes would wear moleskin knee pads as they knelt ramming the sand into the moulds before these were dried in an oven. The metal used was often best-quality Lowmoor iron mixed with scrap. Many manufactures were very co-operative and were happy to share their experience. There is one particular reference to Marshalls receiving a letter from Burrells advocating the addition of one ounce of aluminium to each hundredweight of steel to produce improved castings.

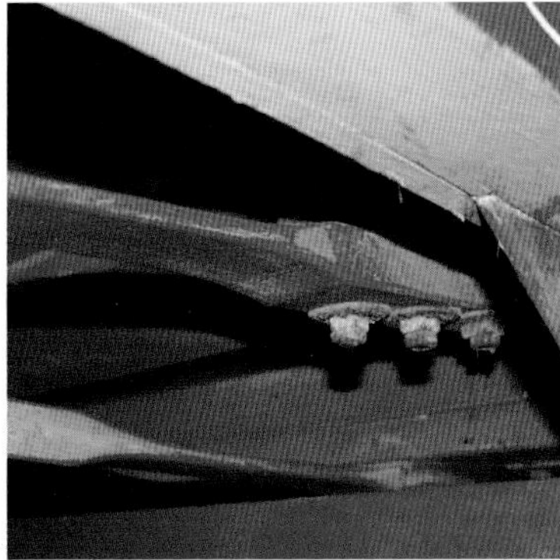

Fig 48: Riddle Crankshaft

Fig 49: Riddle Conrods

Crankshafts

The usual arrangement was to provide three crankshafts, two for the shakers, and one for the riddles. All were bent from solid round steel bar, 1 7/8" for the riddles and a full 2" for the Shakers. Both ran in end bearings mounted externally in brackets bolted into the main framing with the openings in the tongue and groove boarding around the shafts suitably enlarged to allow the shafts to be easily removed once the internal bearings were all split and the external bearing mounting brackets removed. The riddle crankshafts often featured a centre bearing. This centre bearing and all the split pattern big-end bearings were notorious for being difficult to access for lubrication, particularly in the case of the shaker crank big-ends which can often only be reached by climbing onto the cavings riddle and reaching upwards through a removable panel in the grain board. Typically this panel would be around 2' square. With very few exceptions, manufacturers arranged the riddle and shaker crankshafts to be driven at 180 and 220 rpm. It seems surprising that this relatively small difference was perpetuated when it would have been easier and cheaper if they rotated at the same speed.

Drum

Fig 50: Drum

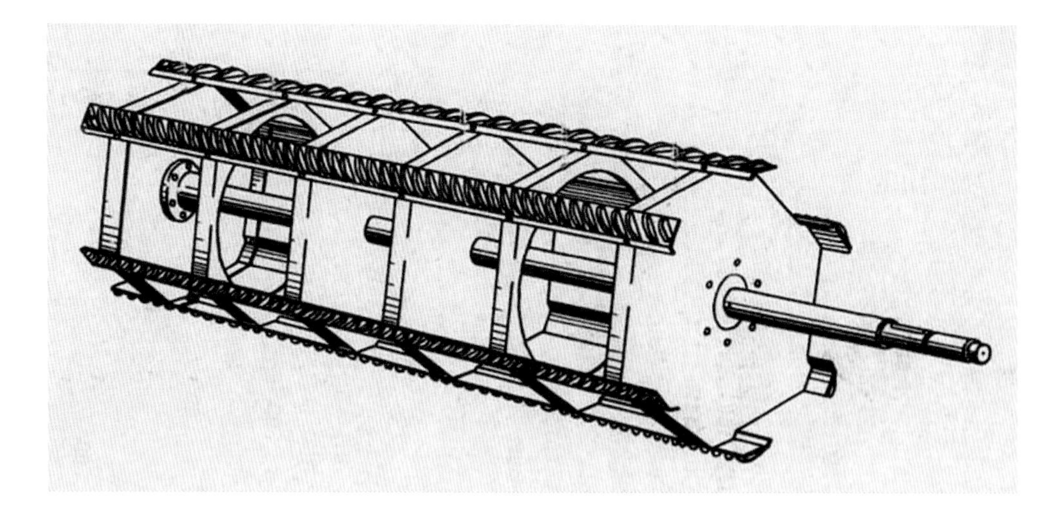

The drum is the full width of the inside of the thresher "box" and is 22" in diameter in most British machines. The diameter of some narrower models was 2" smaller and the larger pattern machines like those intended for export could be 2" larger. Like all the major wearing parts of later machines this would be made of steel. The drum rotated on a simple heavy 2" diameter mainshaft. The most common arrangement would have a number of cast iron bosses keyed onto the shaft, each of these carrying a wrought iron spider sometimes referred to as a drumhead. The beater bars, known in America as "rasp bars", were available in a number of patterns. The most popular on later machines was a fluted section giving "clearing spaces". This fluting was usually at around 30 degrees from the vertical. Many of the bars were designed so that they could be rotated and refitted so as to present a new leading edge after wear had taken place. Because those feeding the drum would be left or right-handed, the sheaves would always be presented with the ears on the same side. This made the wear on the drum very uneven. Many manufacturers designed the bars so that they could be turned end-for-end so that when refitted, the wear could be evened out.

Fig 51: Beater bar - Top

Fig 52: Beater bar - Edge

Those feeding the drum were always advised to keep the flow of material as steady as possible. However well this was done, there would be fluctuations on the load, so manufacturers made the drum deliberately heavy so that it could act as a flywheel keeping the action of the machine as even as possible. Not only were the drums made deliberately heavy, but the weight was concentrated at the periphery where it would be most effective. Because of the extremely high rotational speed of the drum, not to mention its weight, the unit had to be carefully balanced.

Concave

On all but the smallest of machines, the concave is constructed in two halves hinged in the middle.

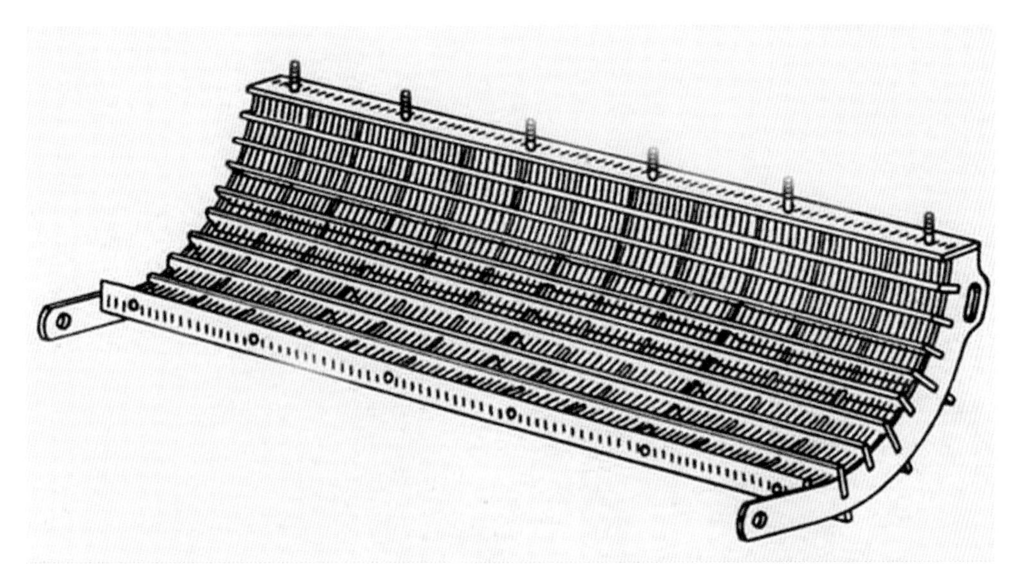

Fig 53: Concave

Together they cover around half of the circumference of the drum, something referred to as wrap angle. Each is made up of a number of metal arcs carrying rows of square section wrought iron bars running parallel to the beaters on the drum. These bars were drilled with wires threaded through the holes. A common wire size would be 5/16" diameter. The metal bars have sharp right-angled leading edges and these become rounded after much use. The bars may be designed in such a way that they can be removed and refitted the other way round to present a new sharp edge. Long rods at the ends of the concave and at the hinge point extend outside the box into pairs of eye-bolts running in mounts on the cast iron brackets that hold the drum shaft bearings. These pairs of bolts can be moved to adjust the clearance between the concave and the drum. On smaller machines, the basic design is similar except that the concave is in one piece without the central hinge and rod thus providing only two adjustment points.

Pulleys

The iron pulleys are sand-cast with any machining restricted to the cambered outer face and the hub centre to take the shaft that often features a keyway. The outer face of the pulley is cambered to provide a degree of "self-centring" for the belt. The width of the pulley is an inch or so wider than the belt for which it was intended, allowing ½" either side. Thus the main drive pulley carrying the 6" belt will be a full 7" wide.

Fig 54: Main Drive Pulley (Garvie)

Fig 55: Main Drive Guide Pulley (C & S)

Shakers

Fig 56: Shaker Bed (Garvie)

The straw shakers, can be variously described as walkers or racks, and in America, Vibrators. These are each around 11' 0" long which represents a length about double that of the periphery of the drum. They are built from pairs of 6" x 1" rails with a large number of cross members with gaps between them that vary between ½" and 1. Most commonly the rails are notched to accept a row of ¾" x ¾" cross battens mounted cornerwise. A few smaller machines use light gauge metal rods as cross-members; these resemble knitting needles. Each shaker might also be fitted with a number of long projecting vertical coach-bolts or diagonally mounted boards to further aid the movement of the straw. Each shaker is framed by 2 pairs of 5" x 2½" timber blocks set 18" apart to accept the cast-iron bearing brackets to the 2 crankshafts. The 2 shafts are constructed to give an equal 2½" throw. The 2 crankshafts are so arranged so that the straw is moved uphill at an angle of around 10 degrees. Most machines have 4 walkers but this number may vary between 3 and 6. Later machines are often fitted with longer shakers which necessitates extending the end of the machine to form an enclosing wind hood.

Check Plates

Fig 57: Shaker Flaps

Fig 58: Shaker Flaps - Adjusting Chain

Immediately beyond the drum, and at one or more points along the shakers, are located a series of items that run the full width of the machine. Variously referred to as doors, or flaps, each of these consist of a row of separate flat metal plates, one per shaker. The plates are attached to a spar hinged from the top of the body and are adjusted by lengthening or shortening a chain running from the spar through a metal button set into the top of the body of the machine. Once the correct clearance has been obtained, a pin is passed through one of the links in the chain to keep it in the same position. The first set of plates are designed to reduce the amount of grain thrown into the straw by the action of the threshing drum. The others, further along the shakers, are designed to slow the travel of the straw and encourage more grain to fall through the gaps between the shaker cross battens. This is described as producing a thin even "mat". Some manufacturers such as A C Tullos of Aberdeen fitted a single curtain to hang from the roof of the wind hood over the ends of the shakers to prevent chaff from being blown about in strong winds. Also from Aberdeen, Garvie fitted a separate curtain to the end of each shaker hanging down around 16" for the same reason.

Shoes

The riddles and sieves are mounted in 2 separate oscillating shoes. Both are framed in 3" x 3" timber and both are operated by the single riddle crankshaft. This has 4 equal throws typically of 1", each with a single connecting rod. The rods are arranged alternately with one pair operating each shoe. As each pair of throws lies at 180 degrees to the other, the speed and movement of each shoe is the same, but in the opposite direction. The connecting rods are of 3" wide x 1½" deep ash and are rigidly attached to the shoes by 3 ½" coach bolts rather than via a pivoting joint. This rigid joint means that the rods are required to flex. To aid this flexure, a section of the rod some 15" long is reduced to 5/8" thick. Typical lengths for the rods are 4' 4" for the top shoe, and 5' 4" for the bottom shoe.

The big ends are each 3" wide and carry cast-iron shoes some 6" long into which the connecting rod ends are notched and bolted with 2 ½" coach screws. This rigid attachment of the connecting rods to the shoes gives a smoother operation and also reduced the number of lubrication points.

Suspensions

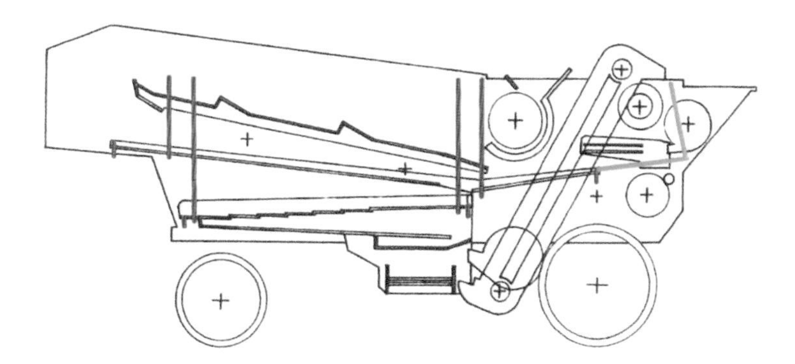

Fig 59: Suspension Slat layout

The 2 oscillating shoes are each suspended on a set of four slats around 5/16" to 3/8" thick and 2" to 2½" wide. These slats are nearly always made of ash, specially selected for of its natural springiness and are often referred to as spring hangers. Lateral movement is controlled by pairs of similar slats mounted horizontally on opposite sides of the machine which act like the panhard rods used to prevent axle-tramp on performance cars. All the slats are rigidly bolted both to the brackets on the machine and to the shoes and so do not pivot but rather flex in operation in the manner as the joints between the connecting rods and the shoes. This gives the 2 advantages of smoother operation and fewer lubrication points. Like the connecting rods, part of the shank of the slat can be slightly reduced in section to aid flexure, whilst still retaining the full section for the last 6" or so at either end to retain strength where the stress is at its greatest.

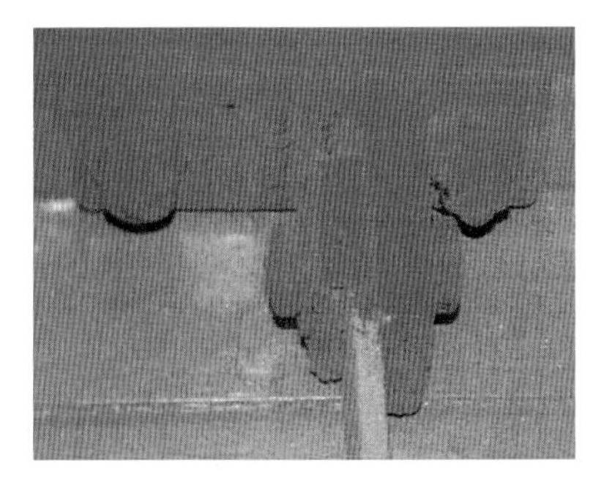

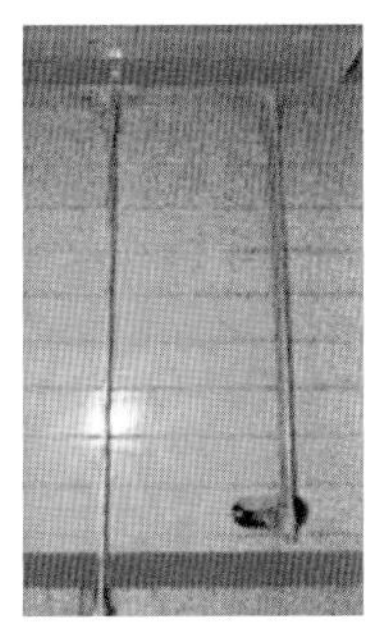

Fig 60: Suspension Slat Top Bracket

Fig 61: Suspension Slats

Fig 62: Slat Bottom Fixing

Riddles & Sieves

Fig 63: Interchangeable Sieve

Fig 64: Cavings Riddle - Wood

Fig 65: Cavings Riddle - Metal

On modern machines, all sieves are flat metal plates with the entire surface perforated with round holes all of which are the same size. The plates of early machines will be of wood. The most popular choice would be a hardwood such as walnut of mahogany but Fosters for one used applewood. For different crops, the interchangeable sieves are available with different sized holes, but the overall design remains the same. Each sieve is framed either in wood, or with a heavy section wire surround and slots into pairs of grooves in the shoes. Some are even fitted with a handle so that they can be removed more easily, like drawers.

The larger or main shoe carries the long stepped sloping cavings sieve on top. This is made up of a number of individual sieve sections as described above. The number can vary from 2 to 7 with a hole size of around 5/8". All the other sieves are made up of a single panel lying to a slight fall, typically around 10 degrees. Below the cavings sieve lie the main sieves. Reading from top to bottom, these are known as:

Top =	Frst, blast, or chob sieve.	Typical hole size: ¾" dia
Middle =	Single primary sieve or pair.	Typical hole sizes: According to crop
Bottom =	Dust sieve.	Typical hole size: 1/8" dia or smaller May be a blank plate.

Manufacturers offered several grades of sieve depending on crop size. For use with cereals, there are 3 principal sizes:
The sieves with the smallest holes are intended for all seeds smaller than wheat.
The sieves with the slightly larger holes are intended for wheat.
The sieves with the larger holes are intended for barley & oats.
Special sieves with even larger holes are for use with peas, beans, etc.

The unwanted light material is blown away by the first blast from the main fan or bottom blower. This first blast is directed at the underside of the chaff sieve. If the machine is designed with a fourth blast, then this is directed at the underside of the grain board below the dressing screens.

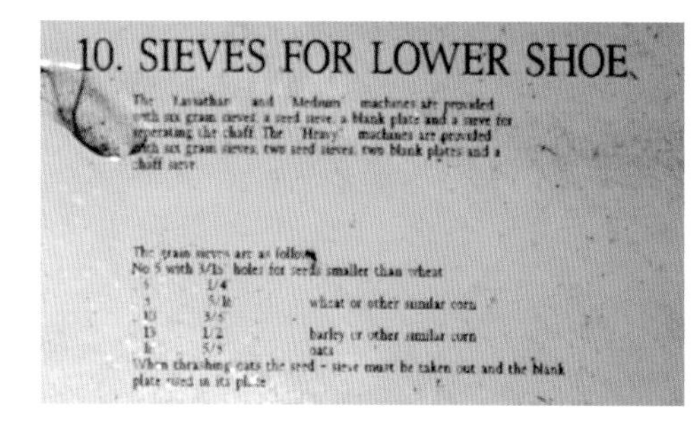

10. SIEVES FOR LOWER SHOE.

The [illegible] and 'Medium' machines are provided with six grain sieves, a seed sieve, a blank plate and a sieve for separating the chaff. The 'Heavy' machines are provided with six grain sieves, two seed sieves, two blank plates and a chaff sieve.

The grain sieves are as follows:
No 5 with 3/16" holes for seeds smaller than wheat
[illegible] 1/4
[illegible] 5/16 wheat or other similar corn
[illegible] 3/8
13 1/2 barley or other similar corn
[illegible] 5/8 oats
When thrashing oats the seed - sieve must be taken out and the blank plate used in its place

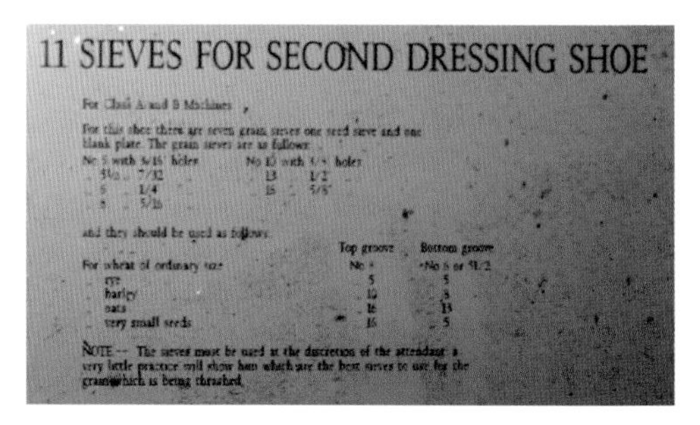

11 SIEVES FOR SECOND DRESSING SHOE

For Class A and B Machines.

For this shoe there are seven grain sieves one seed sieve and one blank plate. The grain sieves are as follows:
No 5 with 5/16" holes [illegible]
[illegible]

and they should be used as follows:

	Top groove	Bottom groove
For wheat of ordinary size	No [illegible]	No [illegible]
rye	5	5
barley	[illegible]	[illegible]
oats	[illegible]	13
very small seeds	[illegible]	5

NOTE — The sieves must be used at the discretion of the attendant; a very little practice will show him which are the best sieves to use for the grain which is being thrashed.

Fig 66: Sieve Sizes (Ransomes) Fans

Fans

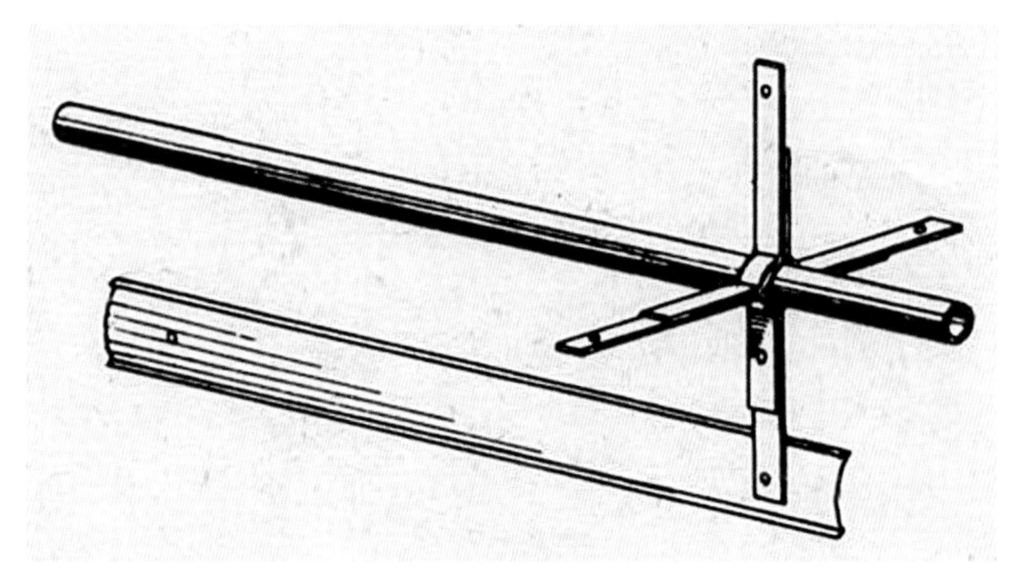

Fig 67: Fan Construction (Claas)

All fans run at between 50% and 90% of drum speed and are designed to produce a blast of air to blow away lighter material, an action known as winnowing. All the fans are constructed in the same way. The ends of a figure-6 section casing are constructed from 1" tongue and groove boarding. The perimeter of the figure-6 is clad a light-gauge metal sheet. The air enters the fan through openings in both end panels. Speed is not adjustable but the blast can be adjusted by pairs of cast-iron doors pivoted at one end and secured at the other with wing nut on a bolt passing through a curved slot

A single central shaft carries pairs of cast spiders carrying 4, 5, or 6 flat blades. The preferred material for the blades is ash. A typical first fan would be 22" diameter, the same as the drum, with blades 6" wide and ½" thick. The first fan can run the full width of the machine but the similarly-constructed dressing fan is much narrower at little less than half the width and of smaller diameter. The blades are around 5" wide and are sometimes made from tinplate.

Fig 68: Bottom Fan (Garvie)

Fig 69: Dressing Fan (Marshall)

Elevators

Fig 70: Elevator Top (Garvie)

Fig 71: Elevator slotted Top Pulley (Marshall)

The grain elevator is usually located on the nearside of the machine, and enclosed in a box of ¾" tongue and groove boarding. On larger machines the elevator is usually, but not always, mounted within the box. All smaller machines, being narrower and more compact, have to have the elevator mounted on the outside. The elevator itself consists of a heavy-duty 5½" wide belt running on a pair of pulleys. The drive puliey has a slotted face to help prevent belt slippage. The grain is carried in a series of pressed metal small cups or buckets each attached to the belt, at around 10" centres, by pairs of rivets or ¼" bolts. Belt drive to the elevator is via a pair of pulleys and is generally taken from the riddle crankshaft. Minor adjustments to the tension of the belt can be made by moving the non-driven shaft in its bearings within an adjustable bracket with a pair of long bolts.

Fig 72: Elevator Inspection Window (C & S)

Fig 73: Elevator adjustment buckles (Garvie)

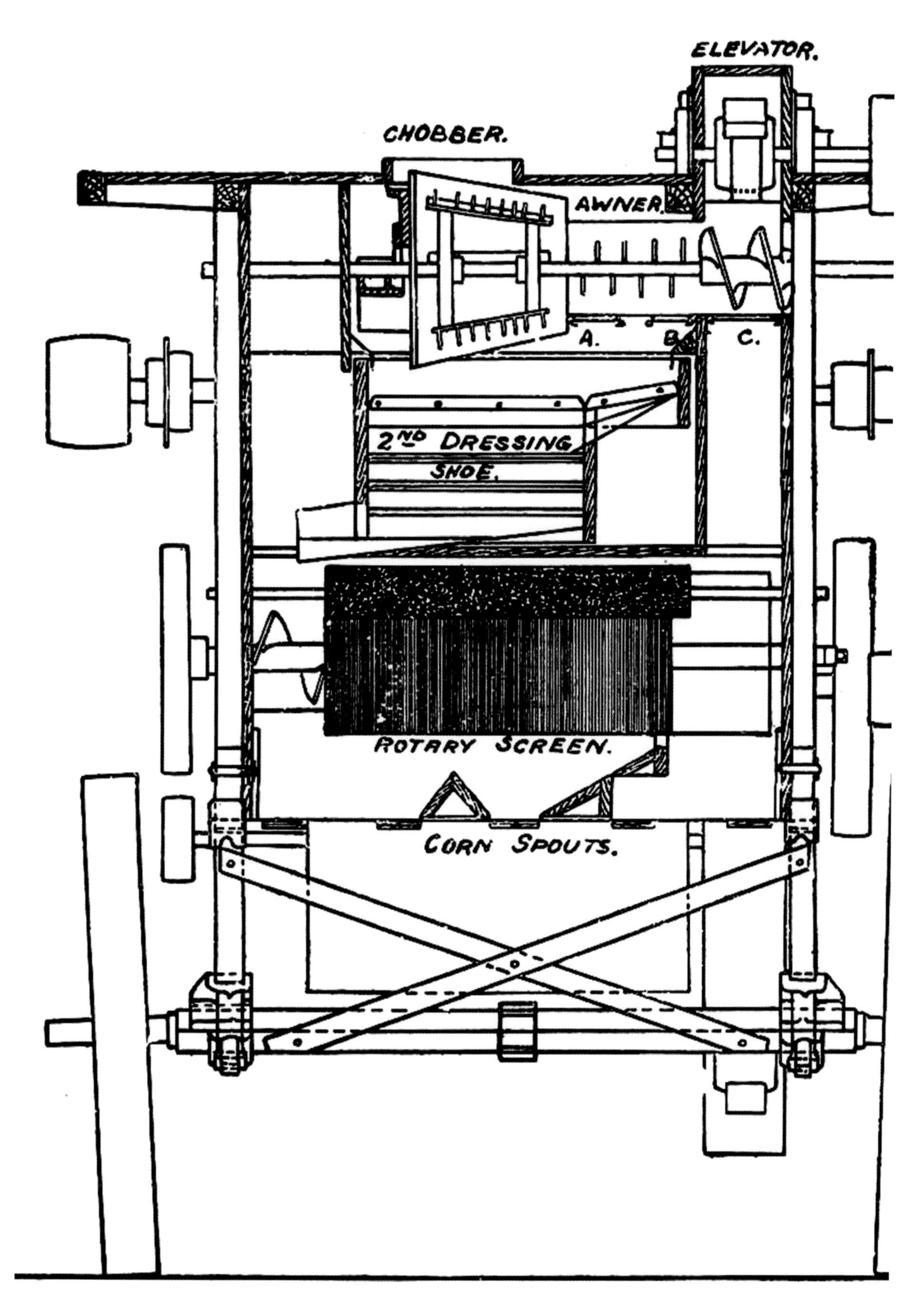

Cross Section

Fig 74: Cross Section

Dressing or Second Treatment

Different parts of the finishing equipment can be bypassed using a set of shutters operated by a number of handles usually mounted externally.

Piler

Fig 75: Piler Casing - Closed (C&S)

Fig 76: Piler casing - Open (C&S)

Perhaps the least understood single component in the entire machine is the set of finishing equipment mounted on the auger shaft collectively known as the piler. Immediately below the grain elevator delivery point lies the short metal auger. On the vast majority of machines, this feeds into a lightweight iron casting comprised of 2 separate chambers each around 18" long, and each with a large inspection panel in the top. The first is the awner, cylindrical with a smooth interior, the second is the smutter, conical with a textured interior. The first is 6" in diameter, the second tapers from 8" to 11" in diameter. The auger shaft turns slightly slower than the drum's mainshaft but still at relatively high speed. If the drum's mainshaft is a little over 2" in diameter, then the auger shaft is slightly less, say 1¾". in the awner chamber, a number of blades with a fairly blunt edge are rigidly mounted equidistantly at 4" centres at 90 degrees to the shaft, and at 90 degrees to one another. Their outer edges clear the casing by about ¼". Here there is no means of adjustment. In the smutter chamber, 3,4, or 6 pairs of brackets are mounted rigidly on the auger shaft. Each pair carries a "knife" parallel to the shaft. The clearance between the knife and the inside of the casing is capable of adjustment through the access panel in the casing. The passage of the grain is controlled by a number of shutters and flaps. The opening and closing of the shutters determines what parts, if any, of the piler are used. The flaps determine how long the grain is subjected to treatment in each chamber.

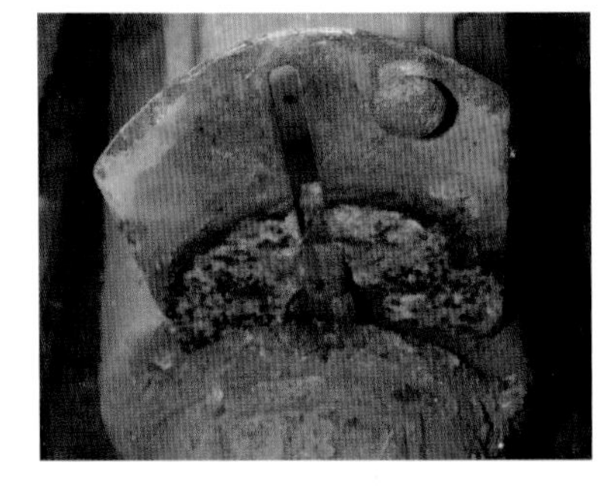

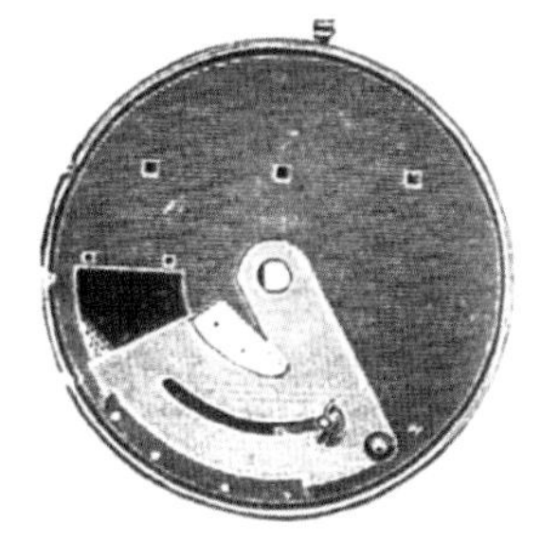

Fig 77: Piler - Awner Shutter (C&S)

Fig 78: Piler - Smutter Shutter (Ransomes) Dressing Shoe

Dressing Shoe

Below the auger shaft with its various components lies the second or dressing shoe sometimes known as the riddlebox. This receives the second blast from the upper or dressing fan. Both components are basically smaller versions of the main shoe and bottom blower but unlike their larger brethren, do not extend the full width of the machine.

Rotary Screen

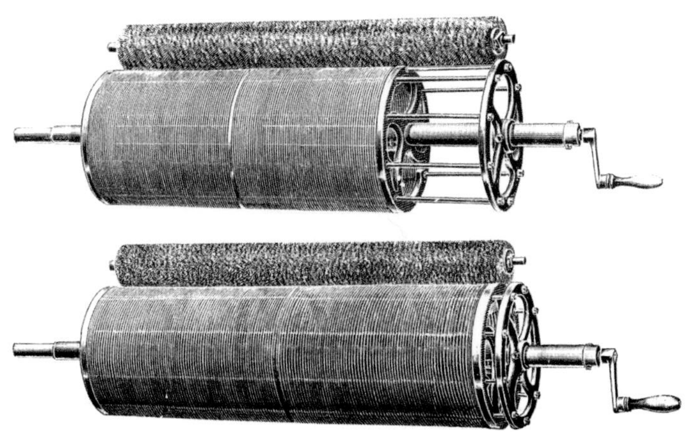

Fig 79: Rotary Screen

Fig 80: Rotary Screen - Interior

Below the dressing shoe lies the rotary screen. By far the most popular manufacturer was the Lincoln company of Penney and Porter. A typical example fitted to a 42" Garvie machine of 1947 is virtually the full width of the machine at 40" and is 14½" in diameter. A single coil of 1/8" wire is wound over 9 bolts 142 times giving a total length of wire of a tenth of a mile. The 9 bolts are of 3/8" diameter and are threaded through a set of 1/8" thick flat metal discs mounted on cast hubs all rotating on a 2" diameter shaft. The 2 discs on the feed end of the shaft are fixed, but the others including the one at the end of the wire can slide along the shaft. Inside the wire is a single spiral of 2" deep spring steel that makes a total of 6 turns. The clearance between the coils can be varied to grade different grain sizes. Average speed for the screen is around 40 rpm. Although this design is dated at 1932, it is very similar indeed to that fitted to a Humphries machine dating from 1875.

Fig 81: Rotary Screen

Fig 82: Rotary Screen - Brush

The gaps between the wires are kept clean by a rotating brush. A row of 2" diameter wooden rollers each 6" long are drilled to take clumps of bristles held in with glue. The holes are 3/8" diameter x ½" deep at ½" and 5/8" centres in 2 staggered rows. This is constructed in exactly the same way as a traditional broom. The overall diameter of the brush when new is 4". The rollers are slipped over a square-section metal axle rotating in a pair of bearings. These can slide in diagonal slots and can be adjusted as the brushes wear.

Corn Spouts

Fig 83: Changeover Shutters on Best Corn Spouts

Fig 84: Sack Hooks on a Rail

Finally below the rotary screen are fitted a row of small hoppers leading to the corn spouts. Each of these can be closed with a vertical sliding shutter. A row of hessian sacks is then hung below the spouts. In order, the spouts are for dust, dust, tail corn, best corn, and best corn. This is the normal layout as read from left to right, but some manufacturers elect to arrange them from right to left. Either way, this relates to the distance between the wires of the rotary screen above. The shutters on the best corn spouts are linked by an inverted T-shaped bar with a handle that closes one shutter as the other opens. This maintains an uninterrupted flow of grain allowing the sacks to be changed without the machine becoming blocked. The limited output from the other three spouts makes this arrangement unnecessary for these other outlets.

Belts

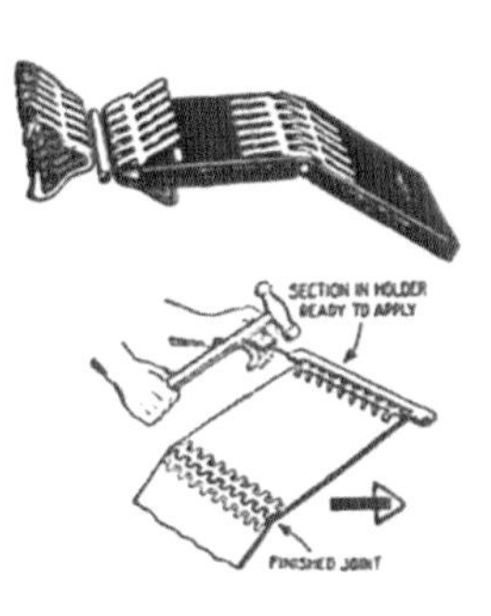

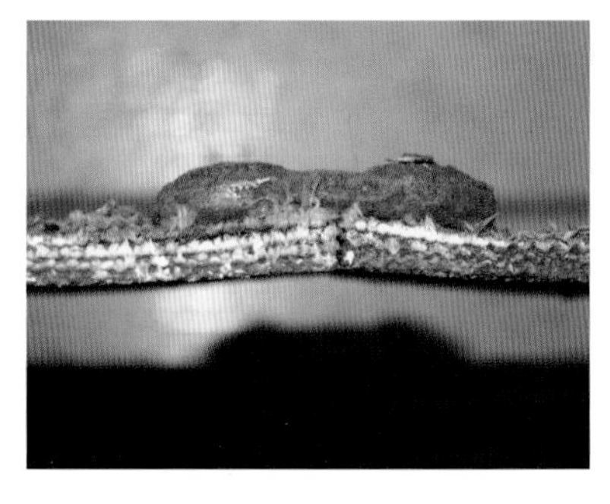

Fig 85: Belt Fastener - 1

Fig 86: Belt Fastener - 2

Fig 87: Belt Fastener - 3

Belts are made either from leather, balata, or rubber. Since leather comes from hides, the length obtainable as a single piece is very limited. This means that any belt will have a number of joints either stitched or glued, or a combination of both. They may also be made up of a number of layers. Balata is a nonelastic natural rubber obtained as a latex from the South American tree Manikara bidentata with properties similar to those of gutta-percha, its processing and uses being essentially the same. It is sometimes referred to as gutta-balata. Rubber is similar. Balata and rubber based belts use cotton duck as the fibre material base and are always made in a number of layers ranging from 3 to 10. A typical main drive belt is 60' to 70' long and 5" to 6" wide. In leather, this would be double layer and cost twice as much as its 6 ply rubber or balata equivalent. The smaller belts are up to 27' long by 1½" to 3" wide and 3 or 4 ply thick. For the purpose of adjustment, all belts have to be fitted with some form of fastener. Popular models included Crocodile, Greene or Jackson Fasteners.

The preferred option for the drive to the governor of the traction engine or portable was a leather belt with Greene pattern fasteners. The choice was based on the fact that this was the easiest pattern to remove and replace. Most machines would have been expected to carry a full set of the smaller belts, but not generally a spare main drive belt.

Sheets

Fig 88: Sheet (Foster)

The base fibre of the sheet would usually be jute or flax, Only the very best quality sheets would have been made from cotton. The gauge of the material would be expressed in terms of the weight in ounces per square yard. Common gauges would be 14, 16, 19, and 22. Most materials were only available in 36" widths, and would therefore require a considerable number of seams. A typical threshing machine cover would be about 21' x 18' with the 2 end laps 4' 6" x 6'. Thus a typical sheet would made from around 48 square yards of material, and weigh around 70 lbs.

Fig 89: Sheet (Garvie)

Chapter Five

VARIATIONS & DETAILS

James Crichton Nameplate

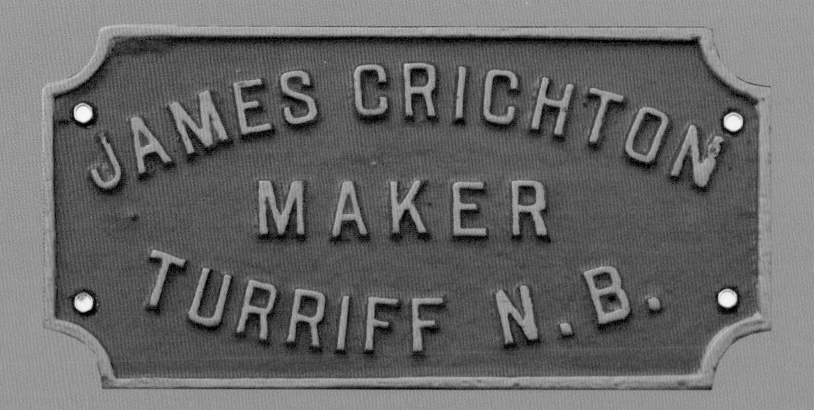

Modern pattern English threshing machines may be divided into 2 distinct groups: contractors' and farmers' models. The 2 are very much the same apart from one important difference: the cheaper farmers' models are minus the final dressing apparatus. Because of this, they are unable to "finish for market". Their output is intended for use only as animal feed and any further dressing would require additional separate barn machinery.

By far the most popular size of machine on the English market was the 54" width version. However manufacturers produced a whole range of sizes. Taking Garretts as an example, their largest model christened the "Mammoth" used 66" wide drum of 24" diameter and required 12 nhp portable or a 7 / 8 nhp traction engine to drive it. Such a machine was capable of handling 90 to 146 bushels an hour. At the other end of the scale, the smallest model that they offered used 48" wide drum of 22" diameter. Marshalls offered an even smaller machine with a 36" wide drum.

Many manufactures also offered their machines in different weights. The most popular option was the "medium" model. The alternative was the "heavy" or "contractor's" model. This was only slightly bigger overall but built more heavily, otherwise the same as its stablemate. In later years, Fosters for one, offered yet another grade of machine, the so-called "Tractor" model. This was made even lighter than the medium model which enabled it to be towed safely by the lighter tractors then appearing on the market. A Marshall catalogue of 1880 lists no fewer than 50 variants of thresher available. Of these 75% were exported.

In general, overall layouts were very similar, but detail design varied a good deal as may be seen in this chapter.

Barclay, Ross, & Tough Nameplate

Box

Fig 90: Early Thresher with angled bottom sills

All early machines, and the majority of modern ones, feature wooden frames, and are also lined in wood. Many earlier designs featured sloping bottom sill members that ran from immediately above the rear axle to the top of the swivelling forecarriage. Later machines are all built with horizontal sills which means that the rear axle has to be held down in extended brackets. This makes for stronger joints for the numerous uprights and easier setting-out during construction. The brackets for the rear axle are generally made from simple flat metal strap bent to a top-hat type section, but Wallis and Steevens for one preferred to fit cast-iron units.

Fig 91: Cast-iron axle brackets (Wallis & Steevens)

The brackets to carry the various shaft bearings were bolted to the upright frame members and better-quality machines rebated the frames to accept these brackets, providing a more positive connection.

Some later machines were built with steel frames. These carried heavy-gauge infill panels that were sufficiently robust to carry other components that would have had to be mounted on the framing of a timber-framed machine. A few also carried cladding panels in light-gauge steel. This was in the style of North American designs, but these would have been only for export. It was claimed that metal-framed machines were less affected by heat and moisture. Shrinkage of wood was then avoided which in turn reduced the requirement to tighten bolts etc mainly in order to prevent excessive wear due to misalignment. This was thought to be especially advantageous if the machines were to be exported to hot climates. Another advantage was that the machines were lighter thus reducing shipping costs. Against this, it was said that steel-framed machines could take on a permanent "set" out of line during shipping. This could cause the bearings on shafts that were out-of-line to wear rapidly. It was also said that wooden framed machines were more easily repaired and that they were more resistant to accidental damage such as might occur if they were to turn over during transport, something that was emphasised in export sales literature.

After the First World War, seasoned timber became increasingly hard to source. Instead of oak and ash, pitch pine was some times substituted. This had been a popular material for use in export machines being not only cheaper, but also lighter which would have reduced shipping costs. Marshalls for one ceased production of timber framed machines altogether. It is interesting to note that during the Second World War, when demand rose for new machines in response to the government's drive for more home-grown crops, their customers forced them to begin building timber framed machines once more.

The frames needed bracing in some way and this was generally achieved in two ways. The simplest method was to rely on the timber cladding. This was more effective when some, or all of it was fitted diagonally. The other approach was to use flat metal straps diagonally to give triangulation to the structure. This was particularly useful to resist the extra stresses resulting from the fitting of additional equipment such as integral trussers. Many designs featured the use of both at the same time. Some machines were designed with diagonal frame members but these were few.

A few machines were clad in sheet metal, but these would be export models in the same style as the North American designs that they would have been competing against.

Roof

Fig 92: Foster Thresher with optional Roof - Closed and Open positions

Some manufacturers offered the option of a sliding roof arrangement as an alternative to sheeting the machine down. Some were of a curved profile, whereas Fosters for one used a pitched design. Generally these roofs were not especially well thought-of because they did little to protect the machine from wind-blown rain and snow, and relatively few were sold.

Wheels

To examine the development of threshing machine wheels is to follow history of agricultural wheel-building in general. Indeed if the wheels are the original ones supplied with the machine, they will give a clear of indication of age of the machine.

Fig 93: Wooden Wheels with Wooden Hub and Cast-iron Hub

The earliest threshers used the same all-wooden wheels that would have been used on all other farm carts and implements of the time. Traditionally these would be built with an elm hub, oak spokes, and elm or ash rim sections referred to as "felloes". All this would be assembled without any nails or screws, the whole thing held together by the iron tyre. This would be heated until red-hot them placed over the assembled wheel using special "tyring tongs". The tyre would cool and contract to tighten the whole structure. Copious amounts of water would be applied, not only to cool the tyre, but also to prevent the wood from catching alight. As the Thresher became larger, so the wheels needed to be stronger and the hub assembly material changed to cast-iron, but the shrunk-on iron tyre and the traditional pattern wooden spokes and felloes remained. A few machines were fitted with the so-called "Stradbroke" composite design of wheel which used a cast hub, wooden felloes, steel tyre, and round steel spokes that could be tightened when required.

Fig 94: All-steel Wheels

All-metal wheels followed. These used a similar cast-iron hub assembly. Each spoke was of flat steel. One end was cast into the hub, the other was riveted to the steel angle or T-section rim which again used a shrunk-on steel tyre. The biggest variation was in the layout of the spokes, which could be arranged either straight or crossed.

Fig 95: Steel Disc-pattern Wheels with rubber tyres - solid and pneumatic

As road surface materials changed, so rubber tyres started to be fitted. These were either solid or pneumatic, but all on pressed steel disc centres. They were often used as replacements for the original wheels. However the change came later than might have otherwise been supposed. The more solid wooden or steel wheels prevented the machine from rocking while in use on the belt. For a better static performance, some would support a machine fitted with pneumatic tyres on jacks, or better, on stands. To keep the machine as steady as possible on the belt, a typical set of pneumatic tyres, say 8.00 x 19, would be inflated to 65 psi. In choosing the preferred wheel, prospective purchasers would need to strike a balance in performance when on the road as against when on the belt. Some builders fitted a set of threaded bars with bottle screws. These were attached to the frame of the machine with eye bolts and carried an angle bracket on the other end that could be slipped over the rim of the wheel and then locked solid by tightening the bolt. A variation of this used a smaller hook bolt to lock the chassis onto the forecarriage.

Bearings

As with wheels, the bearings originally supplied with the machine give a clear indication of age. The bearings on early machines were all plain bushes. The first were made from brass. Some would be of bronze with a high copper content, say 60% to 90%. This then gave way to Babbitted metal, the forerunner of the better-known White-metal bearings that are still in use today. Both use small quantities of copper and antimony, with 80% of cheap lead for the Babbitted metal as against 80% of the much more expensive tin for white metal.

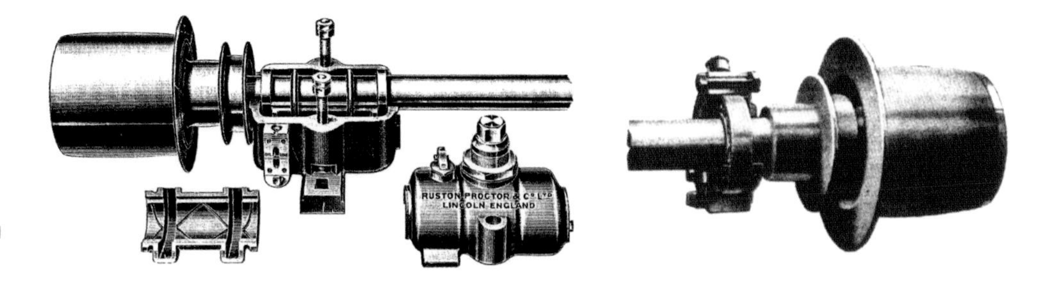

Fig 96: Bearings

Ring Oiler (Ruston)

Roller Bearing (Ruston)

The earlier bearings were wick-fed and required very regular attention. To improve lubrication, the brackets holding these bearings were enlarged to form small sumps holding oil that was conveyed to the bearing surfaces by loose rings on the shaft that dipped into the sumps. These are known as ring oilers. Only the last machines were fitted with ball-bearings and their adoption seems to have been slower than with other forms of machinery. These bearings often featured a knurled knob that could be turned to force the grease between the bearing surfaces. Marshalls claimed that their later machines fitted with ball-bearings throughout, provided a reduction in the power required to drive them of 3 bhp. As has been previously mentioned, problems could occur with bearings if the frame of the thresher went out of alignment. Some manufacturers, Davey-Paxman for one, used a clever design of spherical mount to their bearing brackets to allow some frame flexure without misalignment of the shafts. Incidentally, Davey-Paxman also used an unusual design of mainshaft bearing bracket with a detachable top half which enabled easier removal of the drum assembly, complete with mainshaft, for maintenance purposes.

Crankshafts

All reciprocating parts are given their action by crankshafts and conrods. Generally the shoes and the shakers were driven separately. The commonest arrangement was to drive each shoe as a separate unit. The main shoe, carrying the cavings riddle and all the main sieves, was driven off the riddle crank by one pair of conrods. The grain board, usually extended to carry the dressing sieves, was driven off the same crank by another pair of conrods. The 2 pairs of throws on the crankshaft gave the same stroke but were arranged 180 degrees apart.

The preferred position for the riddle crank was below the drum which made the belt drive off the mainshaft reasonably short, at the expense of greater angulation for the conrod, which were also were quite short. Some manufacturers elected to mount the riddle crank below the shakers. As might be expected, this gave reduced conrod angulation at the expense of an increased drive belt length. Thus a conventional layout might give angulations of 13.6% to 21.2%, whereas on a Garrett machine with long conrods, this reduces to between 8.86% and 14.54%.

This question of angulation is more critical than it might at first appear. Most manufacturers elected to attach the conrods rigidly to the shoes rather than use a pivoted joint. This made for smoother running but also carried the other advantage that it did away with a number of lubrication points, inaccessible ones at that. The disadvantage of the system was that the rod was required to flex along its short length of reduced section.

The shakers were driven separately, by either a single central crank, or more commonly by a pair, each set approximately at quarter distance.

An exception to all these arrangements was Nalder & Nalder's "Simplissima" machine where a single multi-throw crank drove both shoes plus the shakers.

Drum

When considering drum design, it is perhaps useful to consider its two constituent parts separately, namely the Structure and the Beater Bars.

The whole drum assembly is deliberately built heavily so as to act as flywheel. On the most popular size 54" machines, a typical 8 beater drum weighs over 2½ cwt and Ransomes 6 beater drums can weigh over 3½ cwt. The flywheel effect increases if the weight is carried further away from the centre of rotation. Thus a relatively light drum structure could be used effectively if coupled with sufficiently heavy beater bar assemblies at the periphery.

The plain drum mainshaft carries a number of support rings. Most are of a "spider" design. Early models would be a cast-iron centre carrying wrought-iron spokes and rim. In later machines, these would be of steel. Some machines, such as those built by Davey Paxman, featured heavy spiders made entirely of cast-iron. Because of their weight, Davey Paxman used fewer spiders combined with a number of intermediate rings. Some later machines, including those by Garrett, used a plain pressed steel disc design instead of spiders.

The earliest barn threshers used wooden beater bars. Some even constructed the drum itself from wood with only the fittings of iron. The earliest beater bars were of rectangular section but it was soon found that performance improved if the leading edge was angled. This was all the more important because of the relatively slow speeds involved. John Morton, manager of Lord Ducie's Gloucestershire experimental farm, further improved performance by increasing the "angle of attack" of the bars and by chamfering off their trailing edges.

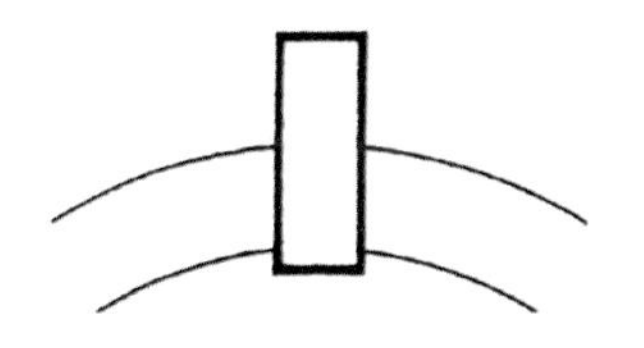
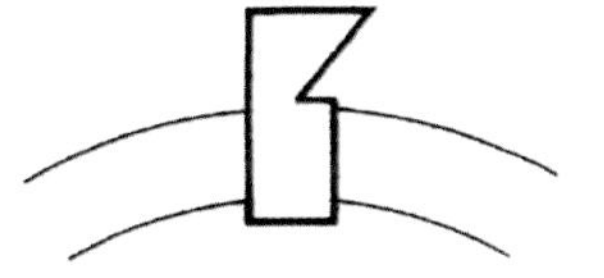

Fig 97 Early Beater Bar Profiles -: Plain

Cutaway

Ducie estate angled

With the advent of steam power and the much higher rotational speeds possible, drum assemblies had to be constructed entirely of metal. However some manufacturers elected to mount the beater bars in a more resilient material rather than attaching them directly to metal, so some Garrett machines used a layer of papier mache in the 135 degree channel section chairs supporting the bars. For the same reason, Davey Paxman used wooden beater-bar beds.

Later beater bars were generally of some sort of ribbed pattern. It had been found that performance was enhanced if the clearance between drum and concave were reduced but some additional spaces were provided for the grain. Garvie for one, used beater bars with the fluting arranged diagonally. These were fitted alternately left- and right-handed so as to move the crop from side to side. In later years, plain pattern bars were usually only found on the smaller cheaper "farmers" machines.

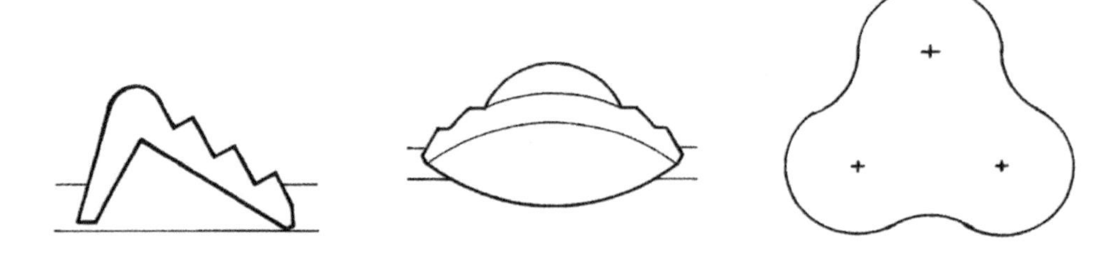

Fig 98: Later Beater Bar Profiles:
Non-reversible

Reversible

Rotatable

Another common design feature was making the bar reversible so as to enable it to present a number of new sharp leading edges. Most were of a "double-edged" pattern but Ransomes for one used a twisted triangular pattern that could provide up to 4 new faces. As has been stated before, manual feeding can be "handed" which means that one end of the bar becomes worn much more quickly than the other. Therefore many designs allowed the bars to be turned "end for end". Some bars were bolted directly into their beds, but many used "drumhead" style hook bolts.

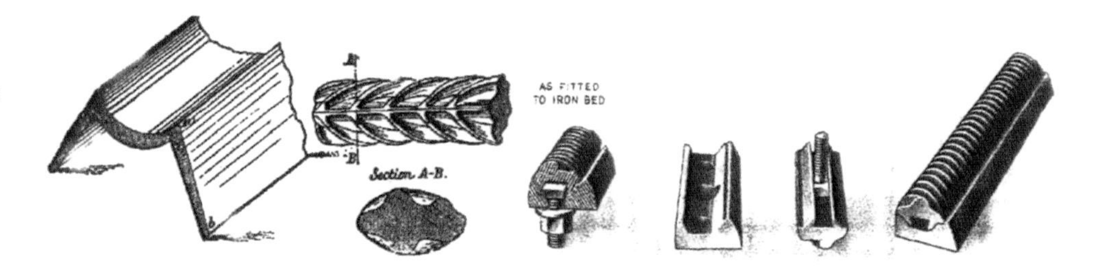

Fig 99: Reversible Beater bars on raised Beds (Garrett) (Ransomes)

In contrast to the above, Humphries used a completely different approach. Like some of the earliest machines such as Ransomes, the drum featured a solid surface with raised bar sections. Despite its antiquated appearance, the "humless" octagonal Humphries drum proved popular with its users and the manufacturers claimed that this type of construction was less prone to flying apart than conventional designs.

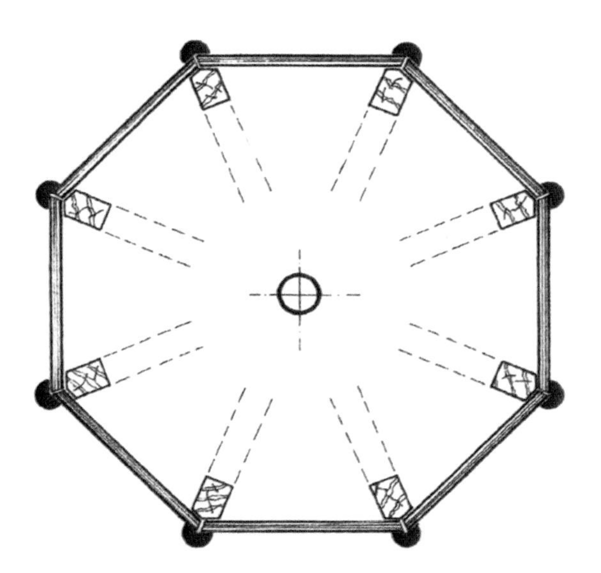

Fig 100: Section of Solid-surface drum (Humphries "humless" pattern)

Concaves

John Ball's design of concave from just after 1800 held a virtual monopoly from soon after its introduction. All larger machines featured a 2 piece unit adjustable via bolts linked to the 3 long bars located at top, middle, and bottom. Most manufactures used eye bolts for all three adjusters but there were some exceptions. Garretts used a bellcrank arrangement for the bottom adjuster, whilst Claytons used a threaded vertical rod on some earlier models. Different manufacturers used varying angles for each of the 2 sections, but the differences are not great, and the total “wrap angle” does not vary much either. In 1928, Marshalls introduced their improved SM design and this featured a novel reversible concave.

A few smaller machines had one-piece concaves adjustable only at top and bottom but these represented a tiny minority of the overall output.

Whilst discussing drums and concaves, it should be remembered that Scotch and American machines were usually of the peg type arrangement. Here the drum is relatively plain with a number of sharp protrusions, the Pegs. Likewise the Concave is also plain with a second set of matching pegs that mesh with the first set. This was very uncommon on English machines because of the damage it did to straw which had such a high market value in the UK.

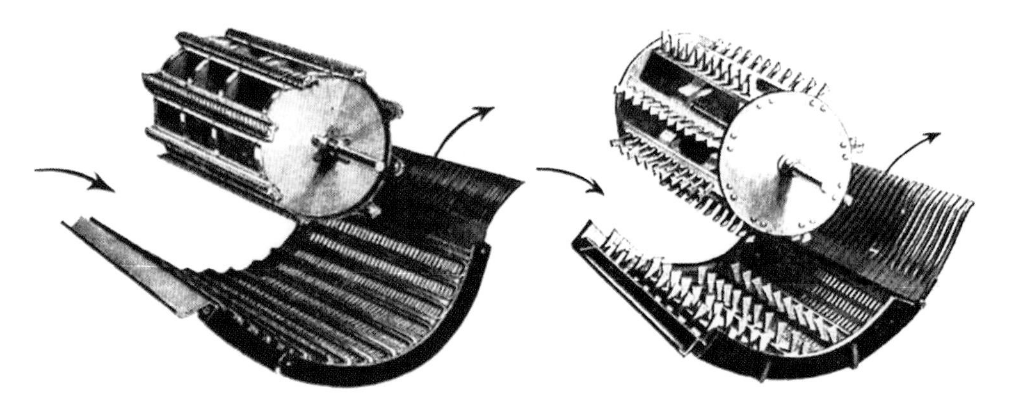

Fig 101: Concaves: English or Plain pattern Scotch or American Peg type with matching Drum

Shakers

When considering shaker design, it is perhaps useful to divide this into profile and action. The frames themselves are basically a pair of wooden rails with a series of slots or holes between them formed by small cross members. They are angled upwards at an angle around 10 degrees and are crank driven to give a reciprocating motion. The vast majority use a series of square-section wooden cross-battens mounted on their corner, but Davey Paxman uniquely fitted perforated metal plates. The length of the shakers was typically around 11' which equates to approximately double the circumference of the drum. Some manufactures elected to increase the length of the shakers by extending them beyond the end of the main frame of the machine. This necessitated a matching extension to the box, the wind hood, and an angled board being fitted under the very end of each shaker to direct grain back towards the end of the grain board. To assist the passage of straw along the shakers, a considerable number of different profiles were adopted. Whilst some relied on a plain flat surface, others used various protrusions to control the flow of the straw. These ranged from simple projecting bolts to angled boards or perforated surfaces, and often in combination. In terms of action, there are two basic systems: Double crank and single crank. By far the most popular of these was the double crank, and even Clayton & Shuttleworth who had been the main advocates of the single crank system, reverted to double crank layout on the last of their machines.

DOUBLE CRANK

Fig 102: Double-crank Shakers

Double crank machines generally had 4 shakers driven by a pair of identical crankshafts with 90 degree throws. The identical cranks meant that the carrying surfaces always remained parallel. Some smaller machine sometimes only had 3 shakers but this was so that standard individual units could be used in a narrower frame.

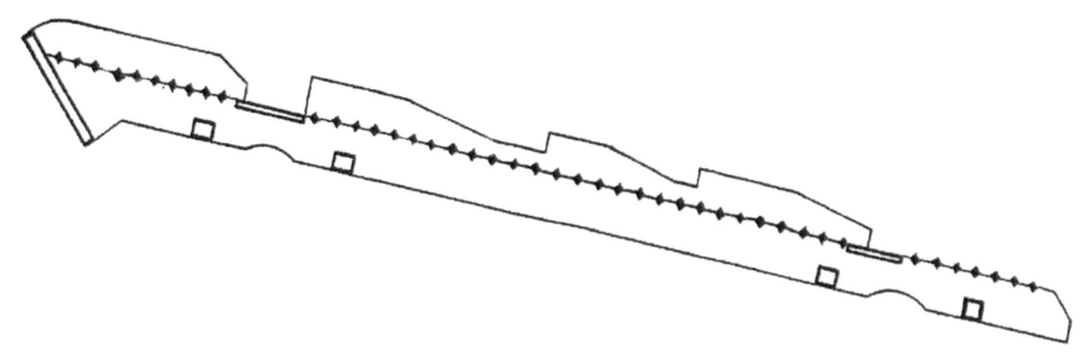

1 *Ransomes*

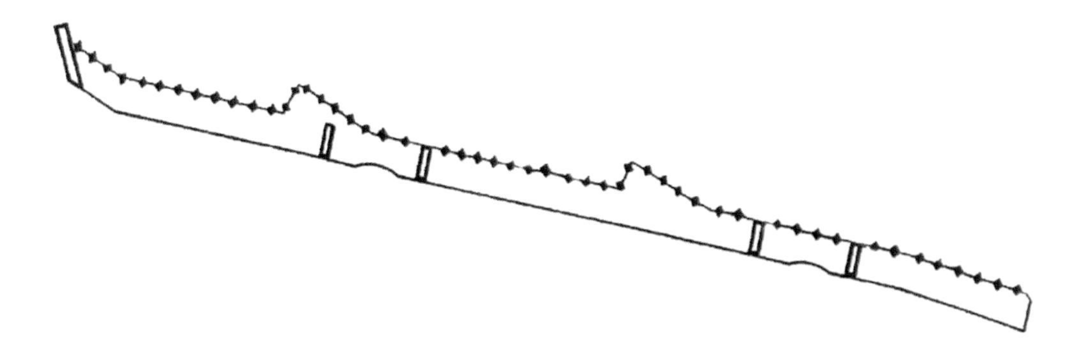

2 *Garrett*

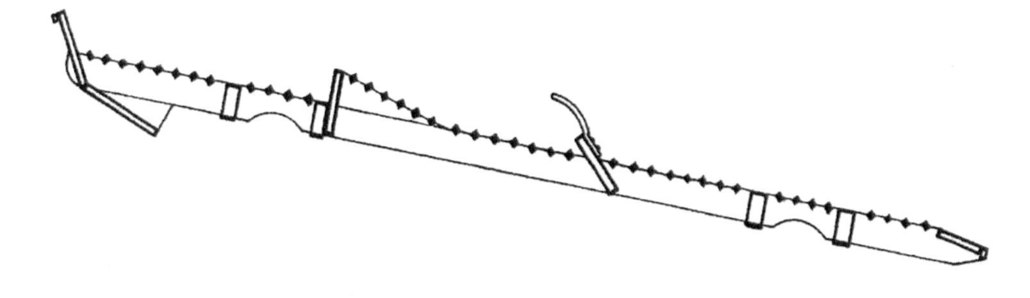

3 *Foster*

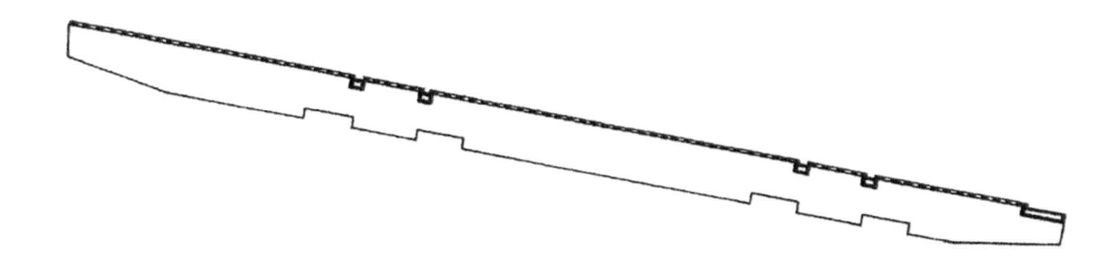

4 *Davey Paxman*

SINGLE CRANK

Single crank machines generally had an odd number of shakers, usually 5, driven by the single crankshaft which was centrally mounted. One end of each of the shakers was attached to the body of the machine by a rocker arm. These were arranged alternately one at the drum end, and the next at the straw outlet end. The odd number was chosen to prevent the flow of straw from being drawn to one side.

Fig 103: Single-crank Shakers

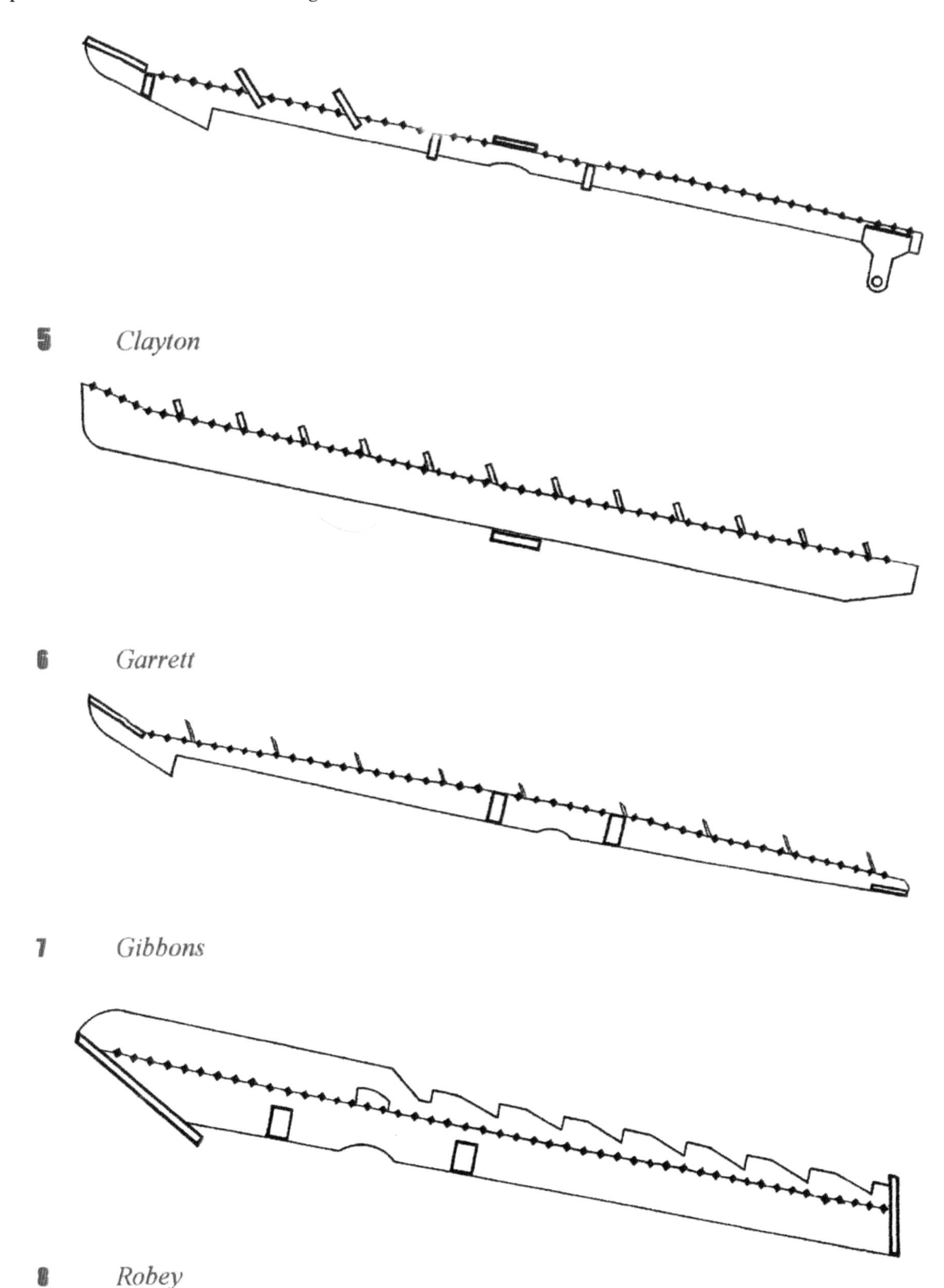

There were a few exceptions to the usual single-crank layout. One was the Humphries machine which used 4 shakers. The crank was mounted at the drum end and all 4 rockers were attached to the other.

Fig 104: Rockers to Shaker Ends (Humphries)

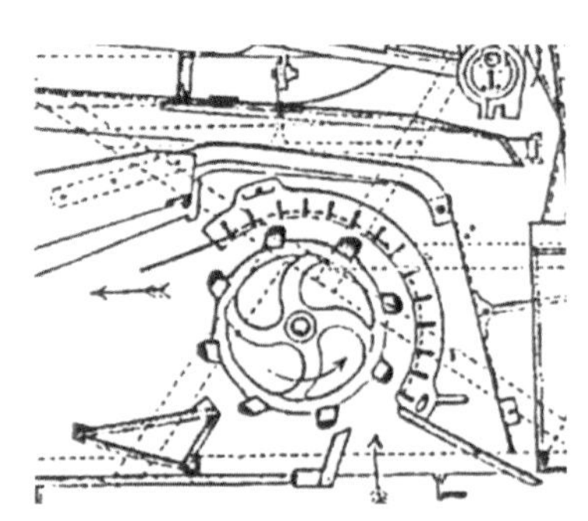

Fig 105: Cranked ends to Shakers (Robey)

Another exception was featured on some of the Robey machines. Here the crank was mounted behind the drum and the shakers were attached by a series of long cranked brackets. It was claimed that this single crank arrangement gave better movement to the straw because of its increased travel but eventually all makers opted for the twin crank layout. As a design guide, it was generally thought that shakers worked best when operating with longer stroke and at a slower speed. A machine built by E R & F Turner featured a single centrally-positioned crank but here the shakers were divided at the centre and hinged at both ends; very few were ever built.

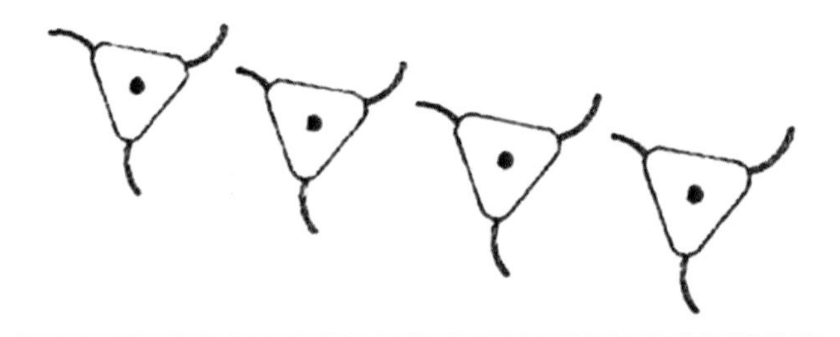

Fig 106: Rotary "Shakers" (Ransomes)

For a number of years Ransomes replaced the reciprocating shakers on a few of their machines, with a series of rotating cylinders with tines providing a "tedding" action rather like those used by the Scotch barn threshers of many years before. The system was not popular, being very prone to chokage, especially when the straw was wet. One thing that made it an attractive option was that it required considerably less power to drive than a reciprocating design. Although this was a virtually unique idea in Britain, it was a more common practice in North America.

Grain Board

The grain board was in two pieces, one running below the shakers, and the other running below the concave. Often this end was extended to collect grain rejected by the second dressing shoe for re-processing. A common arrangement was to arrange the 2 boards as a shallow V with the outlet in the middle to channel the grain down to the main shoe below. The whole unit reciprocated and typically the second dressing shoe was fitted to it using the same drive. A Clayton & Shuttleworth machine has each of the two main bearers for the dressing end of the board to carry a flat metal strap. This extends beyond the rotary screen and is suspended from a loop of leather attached to a bracket above the piler. This is to provide additional support to the dressing screens. A major exception to this layout was employed by Robey who used an apron style conveyor under the conventional shakers. They then used the end of the shakers to drive the dressing shoe.

Cavings Riddles

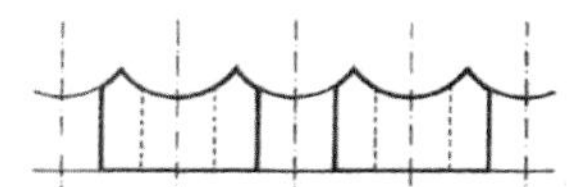

Fig 107: Wooden Cavings Riddle Sections: (Ransomes) (C &S) (Humphries)

Unlike all the other screens on a thresher, cavings riddles were constructed in a number of sections. 5 or 7 is a common number although Garretts made some machines with only 2. Unlike their rivals, the overall length was much reduced. Whereas the cavings riddles on later machines would be exclusively perforated steel or zinc, those on earlier models would often be made of hardwood, mahogany or walnut being preferred. They would be finished with lengthwise grooves, typically 1" wide and ¼" deep. Usually the holes in these boards were simply drilled vertically, typically ½" diameter at 1" centres for cereals, although they were available with holes up to 7/8". The smaller sizes were intended for use with wheat or rye. Some manufacturers wished to push the ends of the cavings back towards the middle of the machine rather than allow them to fall through or worse, block the holes. Clayton & Shuttleworth achieved this by making the top ends of the holes countersunk, and Humphries achieved the same effect by drilling the holes at an angle, the commonest size being 5/8" diameter. The hardwood riddle on 1898 C & S machine is made from boards ½" thick.

Sieves

As with cavings riddles, the sieves on early machines would have been constructed from hardwood. Typically these would be ½" thick and use ½" to 5/8" diameter holes at 1 3/8" pitch. All later sieves are of metal, most of perforated zinc or steel, and a few using woven wire to form a mesh. An exception to this are some of the sieves designed specifically for oats, which can be of a parallel wire design.

In terms of layout, some manufacturers, Ransomes for one, incorporated an additional small sieve before winnowing took place. Also if an integral chaff blower was fitted, the layout of sieves had to be re-arranged so that all the waste material could be led to a single outlet.

Fig 108: Bottom Shoe (C & S)

Winnowing

The basic layout is so designed to allow the blast from the fan to blow away light material whilst keeping the grain on the board or screen. Essentially, this was achieved in one of two ways. Either the board was flat with a small vertical upstand, or the whole board was gently angled. Both layouts require a means of fine adjustment to ensure optimum performance.

Fans

The design of fans themselves was very standardised, but the arrangement of the various blasts was the subject of considerable variation. The fans provide a number of blasts, primarily to winnow the crop, and are numbered in the order in which they appeared on machines. The first blast is provided by the main fan and this is directed to the underside of the chaff sieve. The second blast is provided by the dressing fan and this is directed to the underside of the upper of the two dressing sieves. The third blast is directed at the underside of the caving riddle, more to prevent blockages than to perform a winnowing operation. There may even be a fourth blast. This is directed at the grain board beyond the dressing shoe, and like the third blast, this is primarily to prevent blockages.

The simplest form of popular "farmers" machine without dressing apparatus used a single fan mounted low down immediately adjacent to the bottom shoe. If the machine was a "contractors" model with dressing apparatus then the second blast would be delivered by a second separate fan mounted immediately adjacent to the top shoe, usually outside the main body of the machine and under a small "lean-to" roof. With a view to reducing the number of moving parts, some manufacturers elected to provide all blasts ducted from a single source.

Fig 109: Single Fan (Garrett)

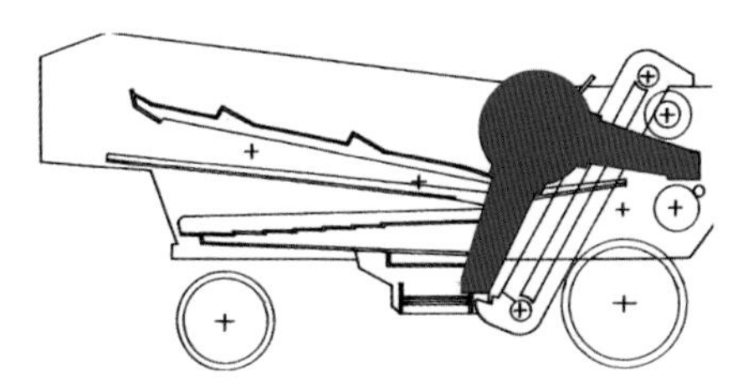

The most popular of these was the Garrett Single Fan model, the subject of a joint patent with thresher shop foreman James Kerridge in 1859. The basic design remained in production for a total of over 50 years with an updated version appearing in 1875. The fan, was a large-diameter narrow-width unit mounted on the drum mainshaft. This arrangement did away with another complete shaft and drive arrangement. At the 1862 Battersea show, one of the judges Harry Evershead, made an interesting comment regarding the life expectancy of threshing machines. In his opinion if a double fan machine could last 8 years, then a single fan machine ought to last 10. But obviously his view was not shared by the majority of manufacturers. The angular wooden trunking on the earlier machines could not have been very efficient and the smoother profiles of the later designs must have given a great improvement in performance. As late as 1904, Prof John Scott in his treatise on barn machinery describes this design as the "most improved of machines".

Nalder & Nalder of Wantage produced a machine of even more radical design where not only was the fan mounted on the drum mainshaft, but all the reciprocating parts were driven off the single shaker crank. This went under the name of "Simplissima".

Fig 110: Single Fan (Marshall)

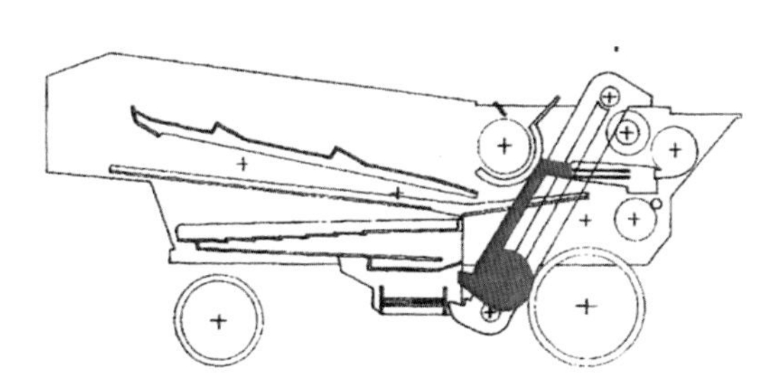

Marshalls built a single fan machine on more conventional lines at the end of the 19th century. Here the second blast was ducted from a normal main fan, but the design did not remain in production for long.

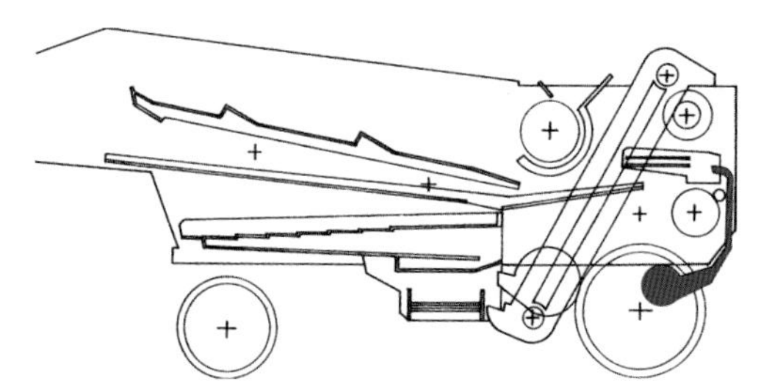

Fig 111: Low-level Dressing Fan (Gibbons)

Gibbons chose to mount their dressing fan within the body of the machine at low level behind the corn spouts with the second blast ducted upwards to the shoe. Later Marshall designs replaced the ducted second blast with a second separate fan. With a view to keeping the number of moving parts to a minimum, the fan was mounted on the awner shaft.

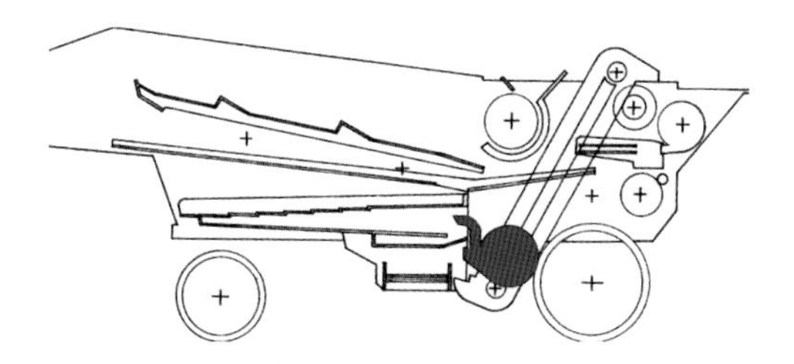

Fig 112: Third "Divided" Blast from Bottom Fan

Some machines featured a third blast directed at the underside of the cavings riddle. This was intended more as a self-cleaning device rather than an additional means of separation. The majority of these were ducted from the nearby main fan and this layout is generally referred to as "divided blast".

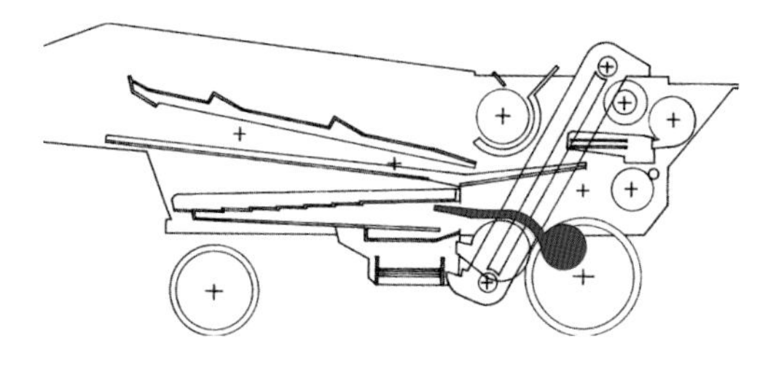

Fig 113: Third Blast from separate Third Fan

Perversely, Garretts who were the main proponents of the single fan system elected to provide this third blast by fitting a third separate fan on some of their machines including the large "Mammoth" model. This fan was mounted at low level behind the main fan which necessitated incorporating a length of ducting.

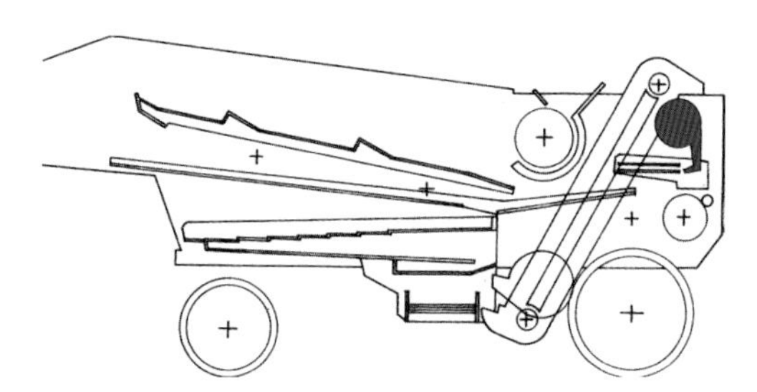

Fig 114: Fourth Blast from separate Fan (Marshall)

Marshalls elected to mount a third fan on the awner shaft. Rather like the self-cleaning action of the third blast, a few manufacturers elected arrange a similar fourth blast arrangement directed at the grain board immediately beyond the dressing shoe. Where these were used, they were all by a "divided blast" arrangement ducted off the dressing fan.

Elevator

On larger machines the elevator is usually, but not always, mounted within the box. All smaller machines, being narrower and more compact, carry the elevator on the outside. Belt drive to the elevator is via a pair of pulleys and is generally taken from the riddle crankshaft. Most machines take the drive to the bottom pulley, but some take it to the top.

Some manufactures replaced the conventional elevator with a blower arrangement. This was far less popular, being much more prone to blockages, particularly in wet conditions.

Piler

The vast majority of pilers are of the cast two-chamber type. The fixed knives in the first awner chamber vary little from machine to machine. By contrast, the knives in the second smutter chamber come in two different patterns. The simpler of these is a straight blade as used by Fosters. The alternative is a multi-tooth design as used by Ransomes. The internal surface of most smutter chambers is of a ribbed pattern, but a few manufacturers, Fosters for one, preferred a pimpled surface.

Fig 115: Smutter Cover - Pimpled pattern (Foster)

The system by which the flow through the instrument is controlled is unique to virtually every machine. Clayton & Shuttleworth use a guillotine-style vertical blade at the end of the awner chamber. This fitted with a wingnut to hold it the desired position. As it is lowered, so the time spent in the awner increases.

Fig 116: Piler grain flow adjustment

Guillotine (C & S)

Rotating (Marshall)

Hinged (Foster)

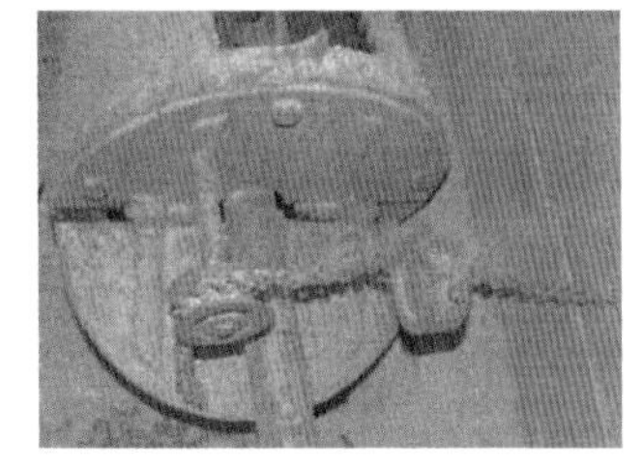

A pivoted plate at the end of the Ransomes smutter acts like a weir overflow; as it is raised so the grain stays longer in the chamber. Clayton & Shuttleworth achieve the same result by means of a plate hinged horizontally across the centre of the end of the chamber.

Fig 117: Garvie Piler: Closed

Open

The Aberdeen manufacturer Garvie used a completely different design. The Piler is a single chamber running virtually the full width of the machine; 40½" long within the 42" frame. Instead of being a casting, the casing is rolled from a single steel sheet. The oval-section blades use a layout similar to that fitted to awners. The only means of adjustment is the flap at the end of the cylinder. Here the operating arm carries a series of notches which enables the speed of flow of the grain through the whole chamber to be varied.

Rotary Screen

The rotary screen is probably the most standardised component on the whole machine. The screens made by Penney and Porter enjoyed a virtual monopoly in later years. The few other designs varied very little from the same basic principle.

Corn Spouts

Fig 118: Patent Sack-holders designs: (Ransomes) (Marshall)

The design of the corn spout arrangement is dictated by the rotary screen described above. As the width of gap between the wires increases, so the size of grain sample increases. This represents the range between dust and best corn. The handing of the screen will mean that some machines read right to left, and some read left to right. The number of intermediate grades may vary but the best corn spouts will always be double because of the volume of material to be collected. Most builders used a simple arrangement of pairs of hooks on a rail to attach the hessian sacks, but Ransomes for one and Marshalls for another, used a more sophisticated clip design.

Chapter Six

ACCESSORIES & EXPORTS

Ruston & Hornsby Trusser nameplate

Self-Feeders

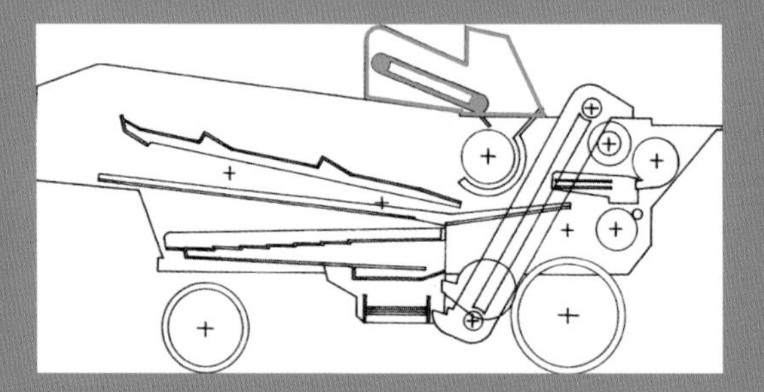

Fig 119: Self-feeder - Schematic layout

Much more popular on larger machines, particularly those intended for export, the self-feeder never achieved the level of acceptance in the UK that it found in North America. The American units had two distinct parts. First there was an automatic band cutter. As might be expected, this component did not appear until the use of sisal twine had completely replaced the earlier wire-tying arrangements. The second part of the machine consisted of a short elevator canvas. UK self-feeders were always the same width as the rest of the machine but this was not the case in the USA. As the crop was always fed "end first", it was found that a wider separator mechanism was required. If an English thresher was described as a 54", an American machine might be a 24 x 36, where the 24" self-feeder would fully load a 24" drum but separation required the greater 36" width.

Fig 120: Self-feeder (Foster Catalogue)

Fig 121: Self-feeder Tine Mechanism

Fig 122: Self-feeder Canvas (Foster)

UK machines used a row of swinging tines at the end of the canvas above the drum, driven by a crank mechanism.

One departure from the almost universal canvas elevator was a design by Gibbons which used a miniature set of single-crank shakers.

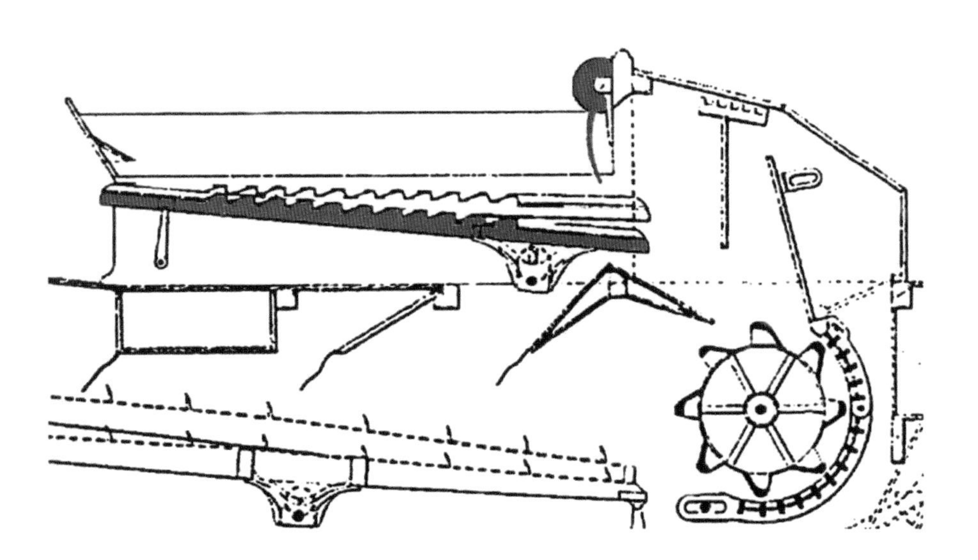
Fig 123: Shaker-type Self-feeder (Gibbons)

Reed Combers

If a conventional thresher is designed to produce fine grain and good straw, then, the purpose of a reed comber is to produce fine straw and good grain. Their use was centred in the West Country where Norfolk-style reed for thatching was not available. The machines were not built by the threshing machine manufacturers, but rather by small local agricultural engineers such as Mounts of Lifton and Murch of Umberleigh, both in Devon. Murch even built a few small threshers. All reed combers are timber framed.

The unit itself sits on top of an unmodifed thresher and is belt driven off the thresher's mainshaft. It is much wider than a thresher, a typical example measuring 72" between the main side members as against the 54" of the conventional thresher on which it stands. This means that the working platform is similarly extended. The crop is carefully fed crosswise with the heads to the left, and it is taken through the machine by pairs of belts that grasp the butt ends of the straw. A series of metal-skinned conical drums, set at right angles to the threshing drum, are fitted with projecting tines that comb the ears off the stalks. Each drum is covered by a light-gauge sheet metal concave rather in the style of the overtype barn threshers of the 1780's, although reed combers didn't make their appearance until the 1920's. A large reciprocating grain board is operated by its own crankshaft which is mounted above the ends of the shakers and belt-driven off the rear shaker crankshaft. This grain board takes the small combed material and gently feeds it to the drum. From there on, the thresher operates completely normally. At the end of the comber's belt feed mechanism, the stalks are released onto an elevator that delivers them to a point beyond the thresher's shakers. As might be expected, trussers were very commonly used in conjunction with these machines, more usually these being of the integral design. A single central cross shaft carries a 24" pulley that is driven by a 6" pulley on the thresher drum mainshaft. This shaft directly drives the sets of feed belt pulleys. These pulleys are typically 8¼" diameter arranged in a lower row of 5 with a row of 4 above. The belts are 5" wide. The combing drums are each driven by a separate 1:1 bevel box off the comber's cross shaft. The drums are around 2' 6" long and taper about 2" along their length. The first drum is the largest with a greater diameter of around 20". Each successive drum is around 2" smaller in diameter than its predecessor. As may be seen, all the shafts rotate at around 250 rpm which gives a feed speed of about 6 mph. The maximum tip speed for the tines on the largest of drums is around 25mph, as against 70 mph for thresher beater bars.

Fig 124: Reed Comber mounted on a Foster thresher

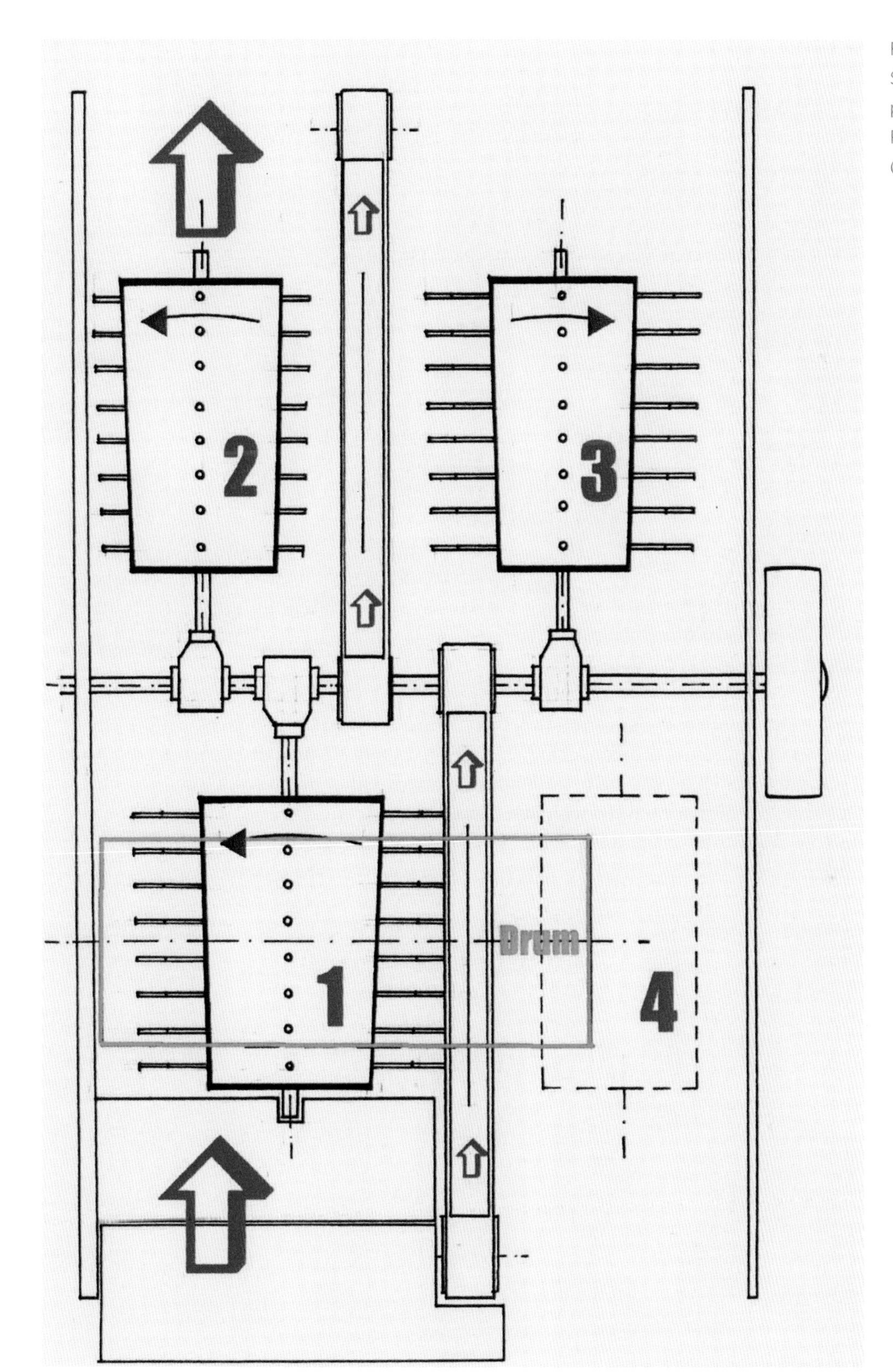

Fig 125: Schematic plan of a Reed Comber (Murch)

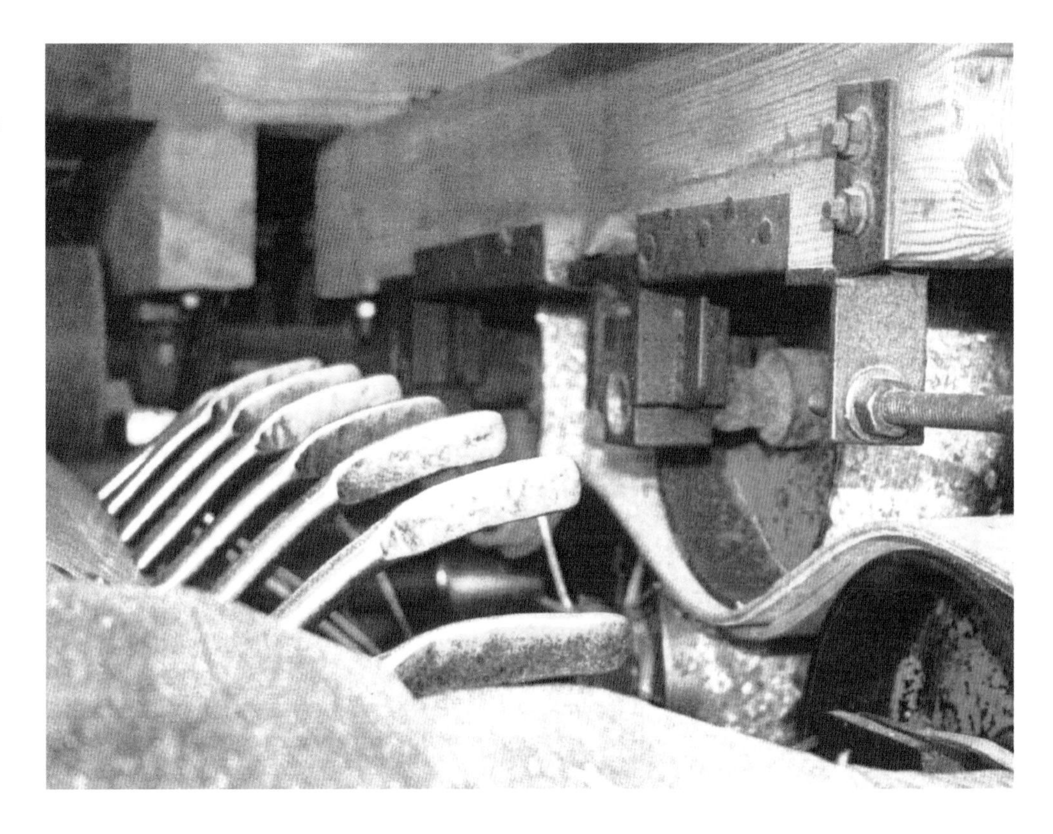
Fig 126: Reed Comber tines and feed belts

The number of drums can vary. The commonest number is 3 as shown overleaf in the illustration. A two-drum machine omits drum number 2. The drums numbered 2 and 4 in a four-drum machine are of a cylindrical pattern rather than conical used elsewhere. Generally the drums are fitted with four rows of 8 tines. These are ½" diameter screwed into the drums with the ends flattened into a shape that resembles a golf club. The tines project some 7" on the primary drums, and more like 4" on the secondary drums.

Fig 127: Reed Comber feeder

Fig 128: Reed Comber drum

Chaff Blowers

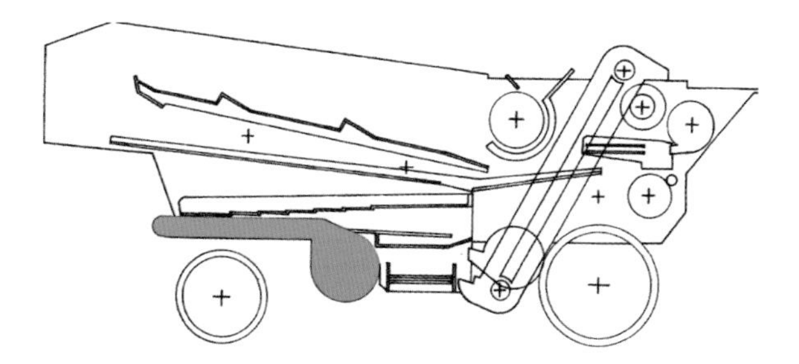

Fig 129: Chaff Blower - Schematic layout

The layout of most machines was arranged to discharge the unwanted material separately from each stage of the sieving process. This could then be collected in a series of baskets placed below the machine. This would give the maximum opportunity to recycle each of these materials individually at the expense of an additional operative with a particularly unpleasant job to perform. Some later machines were fitted with a small integral chaff blower to discharge all this material away from the machine via metal ducting. One of the earliest examples was built by Gibbons in 1881. However this option requires that the overall layout of the machine is modified to bring all the sieved material to a single point. The blower itself has a figure 6 casing and is driven at very high speed off the drum mainshaft. It is mounted low down at the side of the machine in order to collect the material and this location makes it prone to accidental damage. This is a particular problem because the high rotational speed makes the alignment of the unit critical, and hence easy repair less straightforward.

Fig 130: Chaff Blower

Fig 131: Chaff Blower Duct (Garvie) Trussers

Trussers

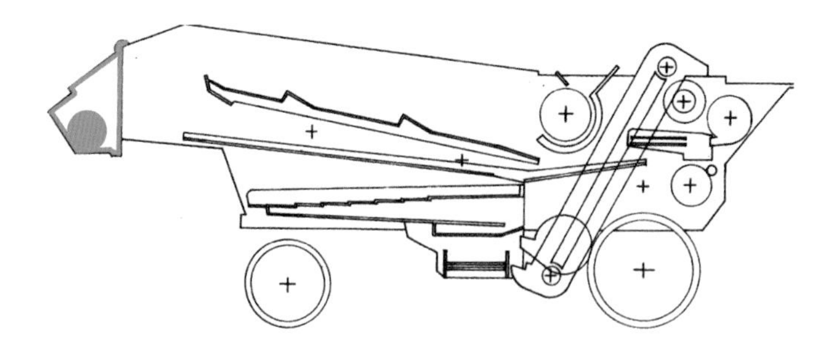

Fig 132: Integral Trusser - Schematic layout

These machines were available in two distinct types. One was a freestanding unit on wheels with a pair of horse shafts, or possibly a towbar. The other was an integral unit mounted at the end of the shakers. Both were driven off the shaker crankshaft and, unlike other components, the drive was always via a chain. This was because the intermittent action of the knotter mechanism would cause even the tightest of belts to slip

Fig 133: Freestanding Trusser (Ruston - Hornsby)

Fig 134: Integral Trusser (Ruston - Hornsby)

The trusser is a derivative of a similar mechanism used on binders. When it is rigidly mounted on the thresher over the ends of the straw walkers, this often requires the addition of metal straps fixed to the wooden frame to carry the extra stresses and weight. The straw is rolled into a long bundle and when this reaches the desired diameter it is tied and ejected. The unit is driven off the shaker crankshaft and rotates at the same speed. Because the flow of straw is not constant, it is not possible to operate the tying mechanism on a simple drive. Rather the bundle size needs to be monitored and the tying triggered at the correct moment making the operation of the machine episodic. The majority of the time there is comparatively little movement but once the tying operation is triggered, a number of parts are required to move in sequence in a very short time and then reset themselves ready for the next bundle. The mechanism requires a good deal of power to complete the operation and this puts a substantial and sudden load on the drive. For this reason these units, unlike anything else on the machine, rely on chain and sprocket drive rather than a belt that could slip. The primary drive from the mainshaft to the shaker crankshaft is still by belt but the fluctuating load has less effect here, partly because of the inertia effect of the reciprocating weight of the shaker mechanism, and partly because the crossed belt means more of the circumference of the pulleys in contact. Unlike binders, most trusser tie the sheaves twice rather than the binder's once using two separate knotter mechanisms. They are also much more susceptible to malfunction because of increased build-up of loose straw and chaff.

THE VARIANT MACHINES

Clover Hullers

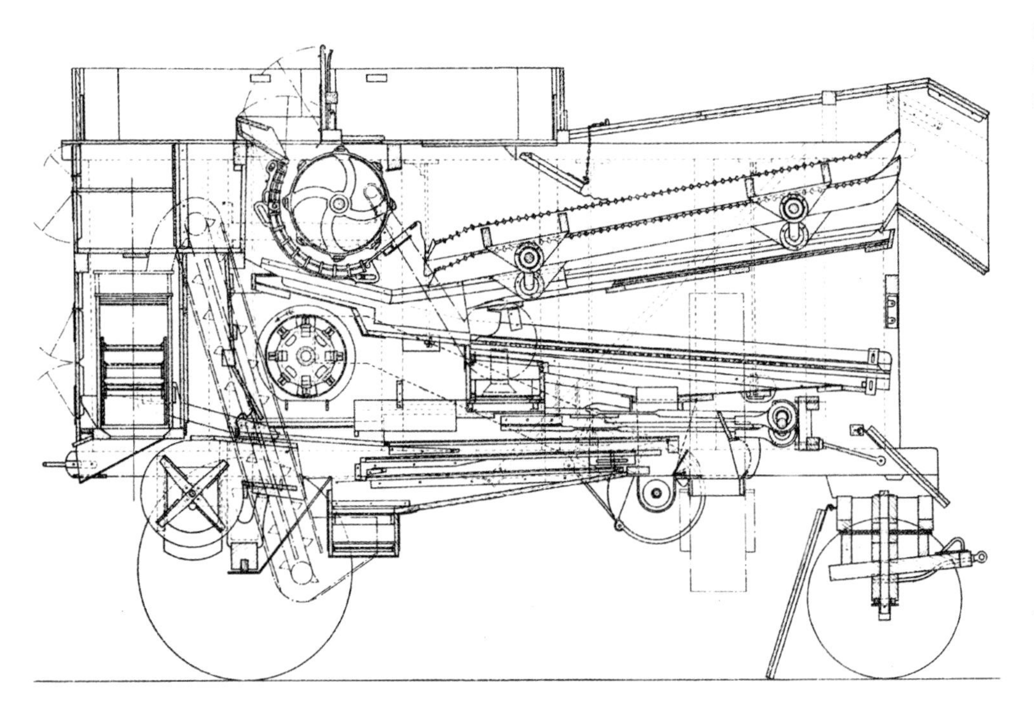

Fig 135: Clover Huller - Section (Ransomes)

A conventional thresher can be used for clover hulling. In order to carry out this operation, the seed has to be passed through the machine twice; once to detach the heads from the stalks, and a second time to remove the seeds from the head. A purpose-built clover huller is a derivative of the conventional threshing machine that features a second drum mounted below the normal drum so that the operation can be carried out at a single pass. The sifting systems are also designed slightly differently to reflect the fact that, apart from fine dust, the material to be collected is of the smallest size.

Fig 136: Clover Huller - removal of second Drum

THE VARIANT MACHINES

Mini Threshers

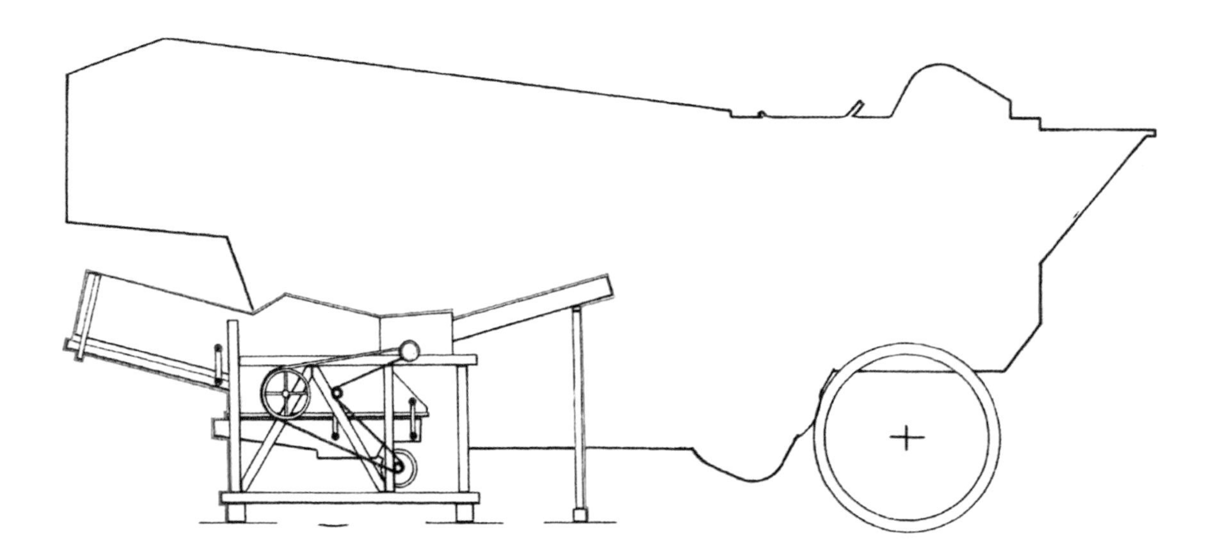

Fig 137: Mini-Thresher - Schematic layout - "Size Matters!"

Above all, these were inexpensive, small and compact. They were produced by local small agricultural engineering companies until comparatively recent times. They were designed to produce small quantities of fodder on a regular basis rather than dealing with a complete harvest and then standing idle for long periods. They were derived from the large early barn threshers, and as the portable machines grew in size, so these fixed machines became smaller. Unfortunately they suffer from many of the defects of early small portable machines. The small diameter of the drum means reduced beater bar speed even when run at the same rpm as larger machines. The narrow width of the drum means crop cannot be fed crosswise. This is bound to cause damage to the straw so that it may be best cut as chaff. Reduced overall size means less opportunity for grain to separate from other material reducing efficiency. The absence of a second dressing facility means that the grain output is only suitable for animal fodder, but this was not seen as a drawback as this was their intended role.

Fig 138: Mini-Thresher - Swedish example

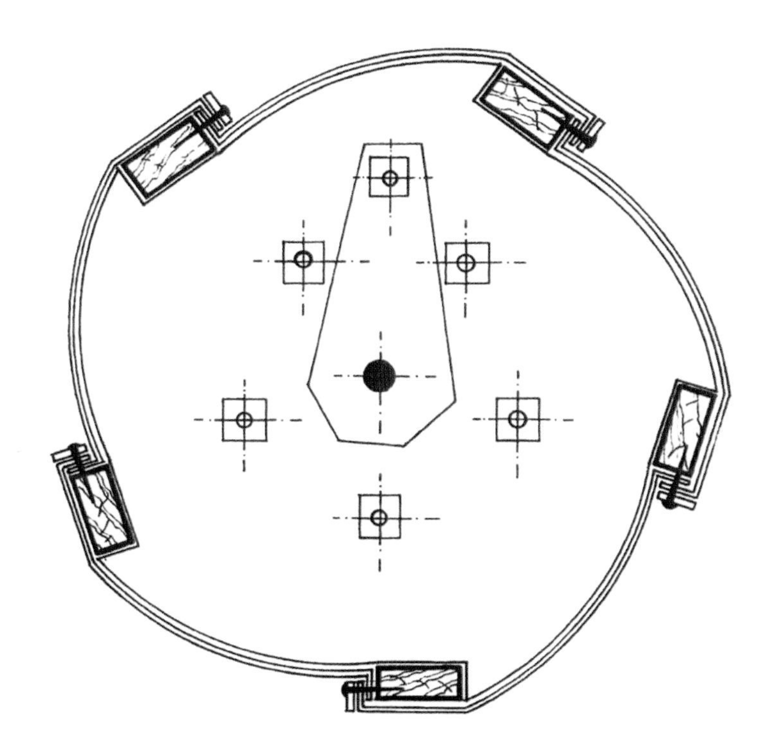

Fig 139: Barn Thresher Drum

Fig 140: Barn Thresher

THE VARIANT MACHINES

The Export Trade

These can be divided into two distinct groups. The first group would be typical English pattern machines with a few minor modifications. The second group would be imitations of the North American machines which were the major rivals in the overseas market. Taking Marshalls as an example, in 1912 exports accounted for 90% of thresher sales. 41% went to British Empire countries, 21% to Europe, 15% to Russia, and 11% to South America. Most of the modifications to the English pattern machines were driven by cost. Traditional oak and ash were often replaced by cheaper pitch-pine. Many machines were supplied with traction engines or a portable as a threshing set. All products destined for export, were supplied in a finish that was very much down to a price. If the machines could be built lighter, then shipping costs could also be reduced. One way of reducing weight was to use steel angle iron for the framing rather than timber. This was also cheaper but caused a problem in early machines. There was a tendency for them to twist during shipment and sometimes the machine would take on a permanent "set" and stay out of alignment. The North American threshers that some British manufacturers sought to imitate were the result of a very different basic approach. The machine was not intended to produce quality straw. This allowed much of the technology to be transferred later into combine harvester design. The overriding parameter was maximisation of grain output, which was limited by straw throughput.

Fig 141: 1902 Case "Agitator" Thresher

Thus a typical North American thresher was constructed in three distinct sections, feeding, processing, and disposal. In the first section, earlier machines used an adjustable hopper to control the passage of the crop into the machine maximising input whilst preventing overloading. Later machines used a self-feeder elevator rather than the hopper. When sisal twine replaced wire ties on sheaves, an integral set of knives could automatically cut the bands increasing speed still further as well as reducing the number of operatives required. A further improvement was to incorporate a governing mechanism which controlled the feeding of the drum. The feeder would not start moving until the cylinder was spinning fast enough and, if the cylinder slowed under the load, the self-feeder would stop making it virtually impossible to jam the cylinder by feeding sheaves too quickly. In the second section, the thresher proper, more commonly referred to on North American machines as the separator, although using the same basic layout, it was significantly different from an English machine. Firstly the drum and concave were of the peg type, so popular on early Scotch threshers. Thus the concave was a solid grooved plate rather than being a mesh of wires. Secondly the drum width was very much narrower as the crop was always fed end-first. Many of the first simple American threshers used drums just 2' 0" wide and were nicknamed *groundhog threshers*. This was not only because of their small size and squat shape, but also because being so light, they vibrated so violently that they had to fixed down to prevent the drive belt being thrown off. The next generation of these machines were nicknamed *bull threshers* because of the roaring sound created by the spinning cylinder. Despite having a drum just 2' 0" wide and 2' 0" in diameter, this type of machine could process 70 bushels of grain per day representing a tenfold increase over the daily output of a single labourer using a flail. In the third section, the straw was taken away by one of two methods. The earlier simpler system was known as the "apron" layout where a series of wooden cross-battens fixed to a pair of chains pulled the straw off a wooden board. The layout was the same as that used on modern manure spreaders and self-unloading trailers. By the mid 1860's, many manufacturers were producing machines based on the "vibrator separator", which had been patented by the Nichols and Shepard Company of Battle Creek, Michigan. In this device, the conveyor "apron" was replaced with a series of straw walkers or racks, connected to an eccentric that caused them to move back and forth longitudinally. Rows of wooden or metal fingers caught the threshed stalks as they left the cylinder, and each progressive sweep shook grain kernels out of the mass, transporting the straw to the rear of the machine. This allowed much more of the grain to be saved than was possible with the previous conveyor systems although these simpler machines were still available. To improve efficiency, the size of the shakers or "vibrators" as they were generally known, was increased. Unlike all other threshers and later, combine harvesters, where all the mechanism is of a single width, the shaker area was made wider so that it was not uncommon for say a 32" wide drum to be mated to a set of 54" wide vibrators.

Fig 142: 1927 Advance Rumelly Thresher with Buchanan Wind Stacker

In America, a popular alternative to an elevator was the "wind stacker" and whist these were rarely if ever used in Britain, a number of British manufacturers offered them as an option on machines destined for export. The original patent had been granted to a Mr Buchanan in 1879 and several thousand of these machines had been sold before 1890. At first, farmers were afraid that the suction effect of the stacker would pull grain away with the straw so Buchanan arranged a convincing demonstration. He had the winnowing blasts turned off then laid a $10 bill beside the grain on the sieves. Needless to say the bill stayed put so that by 1901, Mr A McKein of the Indiana Manufacturing Company that bought the patent rights had sold 9000 machines in less than 10 years, each with the "farmers' friend stacker" logo on the "Happy Farmer" manufacturer's plate. The first North American threshing machine with a metal angle-iron frame clad with sheet-metal panels and was manufactured by the J I Case Plough Works of Racine, Wisconsin in 1904. Despite reservations expressed at first, other manufacturers eventually followed suit. Wooden wheels were replaced by pneumatic tyres on steel discs, and grease cups were replaced by compression grease fittings but there were no further substantial changes to the design of the internal mechanism.

In the early 1920's, the introduction of the small general-purpose tractor, such as the Fordson, created a market for smaller sized threshers designed to be powered by these new machines. Some of these smaller tractors were capable of driving a full-sized machine, but towing such a heavy object with so light a vehicle could prove very dangerous, a problem that still exists for present day-users of threshing machines. By the 1930's, the transition to the combine harvester in America was well on its way. Perversely, some companies continued to produce small manual-fed wooden-framed threshing machines using original 1870's designs. By offering these machines, manufacturers filled an important niche market, in which price and field size would have made the purchase of a large machine or combine untenable.

Chapter Seven

OPERATION

R. Dingle & Sons Nameplate

Viewed in present-day terms, the threshing process requires an inordinate amount of labour. As long ago as 1900, an American publication announced that "The minimum number of hands required in Great Britain are: an engine-driver, a feeder, a sackman, and ten other men to handle the sheaves, straw, chaff and grain. While half as many more again may be needed where the grain has to be carted, as when the thrashing is done in the field in harvest time". This was at a time when still over 40% of the working American population was engaged in agriculture. Large farms would often have their own threshing machines, but the many smaller farmers used contractors to do the threshing for them. The traditional arrangement here was for the farmer to supply the coal, plus the men and horses required to cart water to the engine and corn to the barn. Before the introduction of the Traction Engine, in Cornwall the usual arrangement was for the farmer to organise the horses for the transport of the entire threshing set. The threshing contractor then supplied all the other labour.

Transport

Fig 143: On the Road

In the days of the portable engine, the thresher, engine, elevator, and water cart would have been pulled by teams of horses. Even when the portable first became self-moving and able to tow all the other equipment, one horse was often required to steer the engine. When oil-engined tractors replaced traction engines, the principles of transportation and driving of the thresher remained the same. A train of implements would be towed heaviest first. Generally this meant thresher first, followed by the elevator and the water cart last. However if a stationary baler was in use, this would precede the thresher. Although lighter, balers were more substantially built than threshers and were better able to withstand the stresses of towing another vehicle. Even at a comparatively late date, many of the smaller pieces of equipment were still being supplied fitted with shafts to be drawn by a horse.

Generally the thresher was designed to be towed from the shaker or straw end, but at least one manufacturer, Fosters, offered the alternative of being towed from the corn end. When setting the machine for threshing, it could be pushed into position using a special push-pole. It could also be pulled using the 50 yards of wire rope on the traction engine's winding drum feeding through its 2 pairs of guide rollers. The later oil-engined tractor was often equipped with a rear-mounted winch, sometimes with a large anchor bracket, to carry out the same function. When the machine was being manoeuvred, it was steered by a man holding the towbar. The steering axle was fitted with a pair of check chains. This prevented the axle from turning violently as could happen if one of the wheels were to strike a stone. This was a particular problem before the advent of pneumatic tyres and resulted in a number of injuries to the man steering. The traction engine's detachable rear wheel spuds could be used to "scotch" a wheel to aid the rapid turning of the thresher's pivoting axle. A short chain with hooks was also often carried. Some found that one of the thresher's wheels could be tethered to a gatepost allowing the set to be slewed around a tight corner more easily. All this side-sliding of wheels was only possible in an era that was still using steel tyres on unmade surfaces.

Traction engines especially and, to a lesser extent, their oil-engined successors, were best kept on firm ground as much as possible. Rutting of soft ground made fine adjustments, when turning, impossible as wheels followed any ruts like tramlines and sometimes it was easier to set the thresher to the engine rather than vice-versa. The detritus from previous threshing could often make ground already soft, impossible.

Fig 144: Check Chains

Setting up

When not in use, the thresher was always kept sheeted down. Preferably this would be with a fitted tarpaulin. This was always carefully tied in exactly the same way so that it could be easily removed in the dark, early on a winter's morning, so as to help to make an early start. If it was wet, the sheet would be spread out on the ground to dry. If it was dry, then it would be folded up and stowed straight away on top of the machine. Some chose to cover the top of the machine with straw before fitting the sheet. This created a hump that would throw water off more easily in case of rain. It also meant that the sheet would be less likely to be torn by a machine's sharp corners. For the thresher to operate efficiently, the machine had to be level in both planes and this could be verified using a pair of spirit levels fitted to the frame of the machine expressly for this purpose.

Fig 145: Spirit Level

There was little tolerance permitted in crossways adjustment, but it was found that machines often ran more efficiently when set with the straw end a little higher, especially if the shakers were shorter as they would have been on smaller machines. Levelling the machine was achieved using wedges or levelling blocks on the low side and/or digging a pit under the wheels on the high side. This levelling procedure was made easier if lever lifting chocks, or the heavy 2' 3" long "German jacks" were used. The latter would now be called caravan jacks.

Fig 146: Screw Jacks

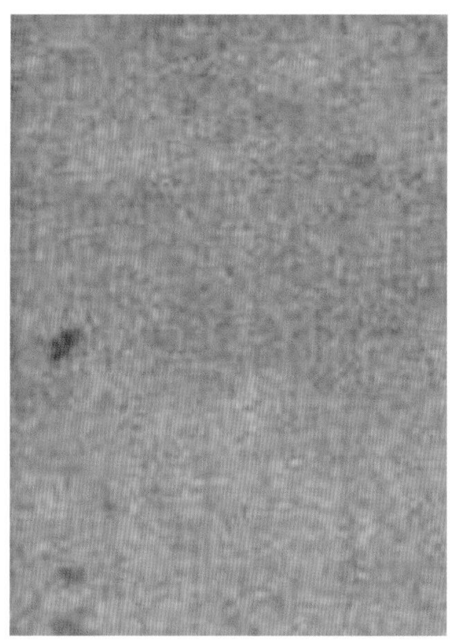

Once level, the machine had to be kept in position. Often this was by using simple chocks but the preferred method was to use chain blocks. These were steel channels 4" to 5" wide facing one another joined by length of chain. Often the traction engine required a jack placing against a front wheel to stop it creeping forward, and this was sometimes used to perform fine adjustments to belt tension.

Fig 147: Chock
Fig 148: Chain Shoe

In addition to being level and not moving forward, the machine had to be prevented from rocking. To help this, large wooden wedges called bolster blocks around 12" deep and tapering were pushed between the pivoting front axle and the frame of the machine. Later machines on pneumatic tyres were even more prone to movement and sometimes had their axles placed on rigid stands. These were constructed of 2½" steel strap on 3' long wooden bases. With the oscillating shakers being moved by a crankshaft turning at around 180 rpm, and the oscillating shoes driven by the riddle crankshaft turning at around 200 rpm, the machine was subject to much vibration and movement, especially during the regular "surges" that occurred whenever both masses moved together. The set ran most efficiently if both engine and thresher were both kept as still as possible. Also this made it much safer, not to say more comfortable, for those standing on the top of the thresher. Some manufacturers fitted screw brackets to lock the wheels in position.

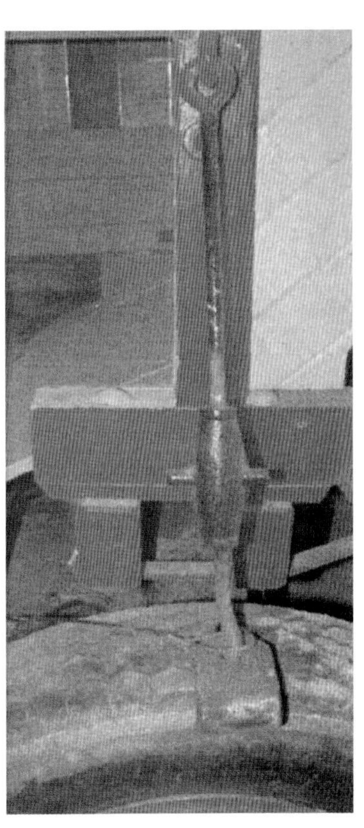

Fig 149: Axle Lock (Tullos)

Fig 150: Wheel Lock (Marshall)

If the smaller secondary drive belts had been removed, these were put on. All of them ran straight with the single exception of the one to the shakers which was always crossed. The 60' long x 6" wide main drive belt was generally kept rolled up when not in use and carried on the cavings riddle. The main drive pulley on thresher drum mainshaft and the flywheel or drive pulley of the source of motive power both had cambered faces. This provided some tolerance when setting up and allowed the set to operate as much as 6" out of line. A useful aid for setting up was a piece of string tied to the thresher and something on the other end to line through with a point on the engine to show when optimum position had been found. A crossed belt meant a greater surface area in contact with the pulleys although this advantage diminished as the length of belt increased. If the power was being provided by a 4 shaft traction engine, wear to the motion gear could be made more equal by running the engine in reverse when on the belt.

If the going was muddy, some operators liked to lay some straw on the ground below the belt so that it would not become dirty if it came off the pulleys. Some preferred to have the wind blowing from the engine to the thresher. This would keep the dust off the engine but directed sparks and smoke to the thresher, and more irritatingly, to those working on top.

Hinged wing boards were fitted to both sides of the machine. These were lifted up to the horizontal position and each propped on a single metal rod like a broomstick with its ends sat in small cast iron brackets. A pair of vertical boards were then slotted into the front and back ends of the extended threshing platform. Another pair of similar boards were then fitted to the sides. One would be mounted horizontally to form an extension foot board on the rick side suspended on short chains from the end boards. The other would be slotted vertically onto the opposite side.

Manufacturers of both the motive power units and the threshers always recommended building up to operating speed gradually. Before work started, many would run the machine at full speed for a short time, with all the fan shutters fully open so that the blasts, operating at maximum force, could clear out any weeds or other detritus left from a previous job. This procedure was often repeated at the end of a day's work for the same reason. In addition to this, all manufacturers were insistent that work should not be attempted until the machine had reached full operating speed. A 1963 Viking combine harvester manual might have been copied the earlier thresher manual word-for-word.

Interestingly, the power required to drive the machine when actually threshing, could be as little as 20% more than when running empty. And referring back to the 1963 Viking combine, if the thresher mechanism requires 8 bhp to drive it, to propel the machine along requires another 19!

Fig 151: St Columb Major c 1890

Feeding

The harvested crop would be delivered to the machine tied in sheaves. The threshing season could last several months so the majority of a season's crop being processed would have been harvested some time before. This means that early in the season, the crop would be coming directly from the field, an arrangement known as "leading and threshing", but later and more commonly, it would be coming from a rick. These ricks were often built in pairs just far enough apart for the thresher to stand between them so that the number of moves for the threshing set could be halved. This also meant that the drum was fed alternately from right and left which evened out wear to the beater bars. The sheaves would be conveyed to the top of the machine by a number of men with pitchforks. This may be seen as the first group of workers.

The sheaves would be passed to the band-cutter, whose job was to cut or untie the strings binding the sheaves. Thomas Hardy's Tess of "Tess of the d'Urbervilles" was employed for this task and the author mentions that this was a not uncommon job for a woman being seen as requiring less physical strength but more dexterity and quickness of hand. If the threshing was being paid as piece-work, the strings were kept as the record of how many sheaves had been processed. The feeder would be standing in the feeding box to take the cut sheaves from the band-cutter and feed them into the drum of the machine. The aim of the feeder was to keep the flow of the crop as even and as continuous as possible. The band-cutter and the feeder may be seen, together with the engineer and perhaps a boy to carry water to the engine, as the second group of workers.

Fig 152: The Drum

Purely from the point of view of grain production, wheat was best fed straight across, barley head first, and oats on the diagonal. In fact wheat was fed slightly butt-first, the weight of the heads causing the stalks to turn fractionally so as to be parallel. However if the straw was to remain unbroken, as would need to be if required for thatching, then the crop needed to be fed straight across regardless. As can be seen, both older and more modern machines required a good deal of skill to operate efficiently and safely. This even feeding was the key to successful threshing. Straw treated more gently was more easily tied into neat trusses, each weighing about 35 lbs, which in turn would be easier to build into neater ricks. An Act of Parliament of 1879 required the fitting of a safety cover to the opening of the drum. These guards were designed to prevent those standing on the platform from stepping into the rotating drum rather than safeguarding the person feeding it. On larger machines a self-feeding apparatus in the form of a small elevator could be fitted. This kept the hands of the person feeding away from the drum and they were sometimes fitted to smaller machines to act purely as a safety feature as an alternative to the conventional drum guard.

Collection

Fig 153: Sack Hooks

Hessian sacks were hung below the row of corn spouts. Usually the sacks were simply hung on pairs of hooks threaded on a rail, but some manufactures, Ransomes and Marshalls to name but two, fitted a more sophisticated spring-clip arrangement. Best grain was collected from two of these corn spouts which were controlled by pair of sliding shutters. These operated in such a way that when one was opened, the other was closed. This enabled the men taking the corn away to change sacks. (a sacks change operation,) without affecting the flow of grain. This was heavy work. A sack of oats weighs 1¾ cwt, Barley 2 cwt, and Wheat a whopping 2½ cwt. Often a handle-wound lifter was used to raise the sacks so that they could be placed on the sackman's back. As the tail corn and dust collected from the other spouts accumulated much more slowly, these other bags did not require frequent changing, which meant that the shutters for these could be of a simpler individual pattern.

Fig 154: Scene from a Clayton & Shuttleworth Brochure c 1900

At the other end of the machine, the straw was delivered by the shakers. Most commonly, it was allowed to fall to ground. A second option would be to use a trusser to tie the straw into bundles called unsurprisingly, trusses. This machine could either be a free-standing unit with a hopper feed and fitted with a towbar or more commonly, a pair of shafts for a horse. Alternatively, it could be an integral unit mounted at the end of the machine immediately over the shakers. Either way, it would be driven off the shaker crank by a chain rather than by belt. This was to prevent the machine stalling when subjected to the intermittent load of the knotter action. Whether the straw was loose or in trusses, it could then be built into a rick. The building of this was often aided by the use of a straw elevator fitted with a large hopper, and driven by belt off a pulley on the rear walker crankshaft of the thresher. The building of the rick was often entrusted to older workers since this task was thought to require more skill and perhaps a little less physical strength.

A second option would be the use of a chaff cutter also driven off the thresher shaker pulley but this time by a belt. The chaff could be collected in special long hessian sacks hung from a pair of spouts operated by a similar double shutter arrangement fitted to the best corn spouts. Alternatively a chaff blower could be used. This took a secondary belt drive off the chaff cutter and blew the chaff through a series of metal tubes. A more recent third alternative would be the use of a stationary baler but this would more likely be a stand-alone unit, requiring a separate power source.

Fig 155: Threshing at Trevean, Gulval, 1952

The third group of workers were those who dealt with the various separated parts of the crop. The best grain was taken away by the sackmen or waggoners. The straw was passed to the rick builders. Alternatively the cut chaff could be carried away by other sackmen. Someone would always have the unenviable task of removing the other detritus such as dust, seeds, tail-corn etc from below the machine, where it often fell into baskets. An alternative to this arrangement on more modern machines was an integral chaff blower. This conveyed this small-sized material away from the thresher through a set of metal tubes and was a smaller version of the freestanding chaff blower that could be used in conjunction with the chaff cutter. If the manufacturer offered an integral chaff blower as an option, then the overall layout of the machine had to be modified slightly so that the various outlets for the unwanted material were all brought to a single point.

Fig 156. Threshing Demonstration at St Agnes 2005

Maintenance

Fig 157: Lubrication Plate (Garvie)

The bearings on threshing machines fall into 3 distinct groups. The earliest machines were equipped with wick-fed plain metal bushes. These required attention at least twice a day, usually before the start of work and again at midday. These were replaced by ring oilers. These have a greater oil capacity with a loose ring on the shaft taking oil from a small "sump". These require attention only around once a week. Ball and roller bearings appeared as an option in the 1920's and became increasingly popular. Later versions of these featured a screw-operated pressure feed for grease, which meant that they had to be inspected even less frequently.

Because if its high rotational speed of nearly 70 mph, the drum had to be kept in perfect balance. Rebalancing a drum was a fairly major operation, which required its removal. This usually had to be carried out as follows: First of all, the drum safety guard was removed. The top bar of the concave was then removed allowing the top half of the concave to pivot on the centre bar and fall back away from the drum. The adjusters could be left in their original positions whilst this was carried out. Next, the pulleys on one end of the mainshaft were removed, care being taken not to damage the keyways. The keys were then removed from drum spiders or discs; an awkward job that meant reaching down through the drum. At this point ropes were passed through the drum and secured to support its not insubstantial bulk, some 2 to 3 hundredweight. The mainshaft was then withdrawn through the bearings, allowing the drum to be lifted out. This was easier with a block-and-tackle, but possible just roped to a beam, if the maintenance team was feeling strong enough. The mainshaft was then re-inserted into the drum and some of the keys re-fitted. Some also elected to re-fit the removed pulleys. Some manufacturers speeded up this process by making a simple alteration to the box. A short section of each top rail immediately above the mainshaft was made removable. A piece 6" long was morticed in and held by a flat metal strap on its underside secured with coach bolts. Between this and the mainshaft, the timber cladding was replaced by a small tinplate panel. With these removed, the drum could be lifted out complete with shaft and pulleys. The disadvantage was that the structural strength of the box was somewhat reduced.

Whichever design had been used, the drum could then be balanced by rolling the mainshaft along a pair of "knife edges". These needed to be roughly parallel and set a few inches further apart than the width of the drum. These didn't need to be absolutely parallel and the mainshaft didn't need to be absolutely horizontal, but these knife edges HAD to be perfectly level. Some thresher manufactures offered an option where a pair of these edges could be fitted to the machine on either side of the feed hole. Although the machine would be fitted with spirit-levels, these knife edges were often designed with a set-screw arrangement to give an even finer adjustment for level, the manufacturers feeling that spirit levels alone were inadequate for this task.

Alternatively the operation could be carried out in the workshop but the same degree of care would still be required. The drum would be rolled along the knife edges on the mainshaft and, if perfectly balanced, would come to rest in any position. If one side was heavy then the drum would tend to return to one position. Any adjustment would be made by the addition or removal of balance washers or small weights. A common test was to place a penny on any of the blades. This would be sufficient to detect any imbalance. In view of the high rotational speed, extreme care had to be taken to ensure that nothing became detached.

Before the drum was replaced, it was prudent to check the concave and carry out any repairs, such as straightening any wires bent by stones. This would not be possible with the drum in place. The edges of the cross bars would wear in time and some concaves were so designed so that they could be removed, stripped, and then reassembled with the bars reversed, thus improving performance by presenting a new keener edge. If this procedure had already been carried out, it was possible to plane the bars to give better edges.

Refitting of the drum was simply the reverse of the removal process, with special care being taken so as not to knock the drum, which might throw it out of balance again.

The main reason why the drum might require rebalancing would be damage resulting from a foreign body falling into the mechanism. Another reason would be if alternate bars had been removed, as was recommended when dealing with certain larger crops such as beans. This attention to detail regarding bar replacement continues to this day. John Deere for one sell replacement beater bars in pairs. The intention is that the new bar replacing the worn item is balanced by a second new bar fitted to the other side of the drum 180 degrees away.

The tension of the main elevator belt was adjusted by sliding the idler pulley, together with its shaft, bearings, and mounts, usually along pairs of long threaded bars and held firm by pairs of lock-nuts. When the full extent of the adjustment had been taken up, the belt would require shortening which could be carried out as follows:

One of the inspection covers was removed. The belt was turned by hand to expose the joiner. The adjusters were slackened off fully and the joiner cut off the belt to produce two clean ends. The ends of the belt were overlapped and more material cut off one end so that the new ends just touched. The new fastener was fitted and the belt tensioned by the adjusters. The inspection panel could then be replaced.

To replace the belt completely, all inspection covers had to be removed and the adjusters fully slackened off. Draw strings were then attached to one of the elevator buckets and the elevator turned until the draw strings were fed around the whole unit. The draw strings were untied from the bucket before the belt was cut. The belt could then be withdrawn complete with all the buckets. The buckets were removed from the belt by drilling out the rivets taking care not to enlarge the holes in the buckets themselves. A new belt was laid out beside the old one and holes punched to match the existing. The old buckets were then re-riveted on. One end of the new belt was attached to the draw strings and the whole assembly pulled through. A new fastener was fitted to join the ends. The assembly was turned once more and the draw strings untied. Finally the belt was tensioned by the adjusters and the inspection panels replaced.

Storage

Because of the seasonal nature of the work, storage was an integral part of a threshing machine's life, unlike the motive power units used to drive it that functioned best if used regularly. Some reckoned that a good season would give 100 days between harvest and Xmas with another 50 days after the worst of the winter up to Easter with possibly a few odd jobs even later. In some areas, the season could last almost the full year with just a short break for maintenance and repairs.

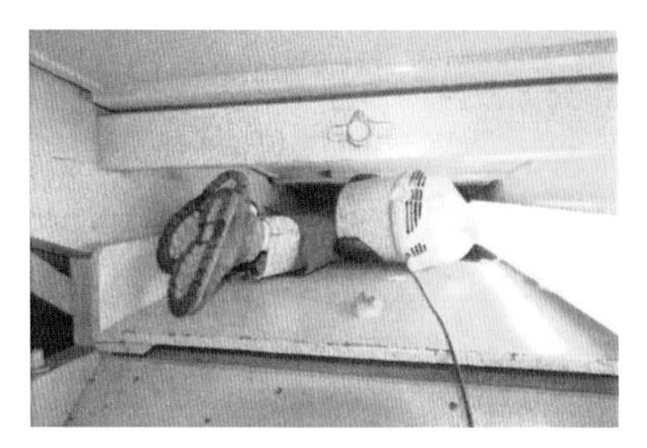

Fig 158: End-of-season Clean

The first thing to do after the season would be to thoroughly clean the machine. This was primarily to stop vermin infestation, but leaving the vegetable material trapped in the machine would also lead to corrosion of all metal parts. This is not only because of the inherently acidic nature of the trapped plant material, but also its ability to expand and retain moisture. If left for lengthy periods, the grain has even been known to germinate. If the machine was fitted with pneumatic tyres, these would deteriorate when the thresher was left standing for long periods, so it was best if the machine was placed on stands. These could well be the same stands as those used when threshing. A number of the wearing parts might require replacement such as the brush to the rotary screen. Many of the bearings, particularly the earlier designs, were constructed in such a way as to allow tightening and adjustment.

Should a re-paint be required, a favourite colour scheme would be deep red for the chassis and beams, and salmon pink or orange for the boarding. The reason for this choice of colours is simple. The favoured wood preservatives at the time all contained high concentrations of red lead, and to have paints of other colours would require large amounts of additional pigment, which would increase cost. Finally, the sheet could be treated with composition or dubbin before being put back on the machine.

Fig 159: Starting Again

Table 3: Relative Speech

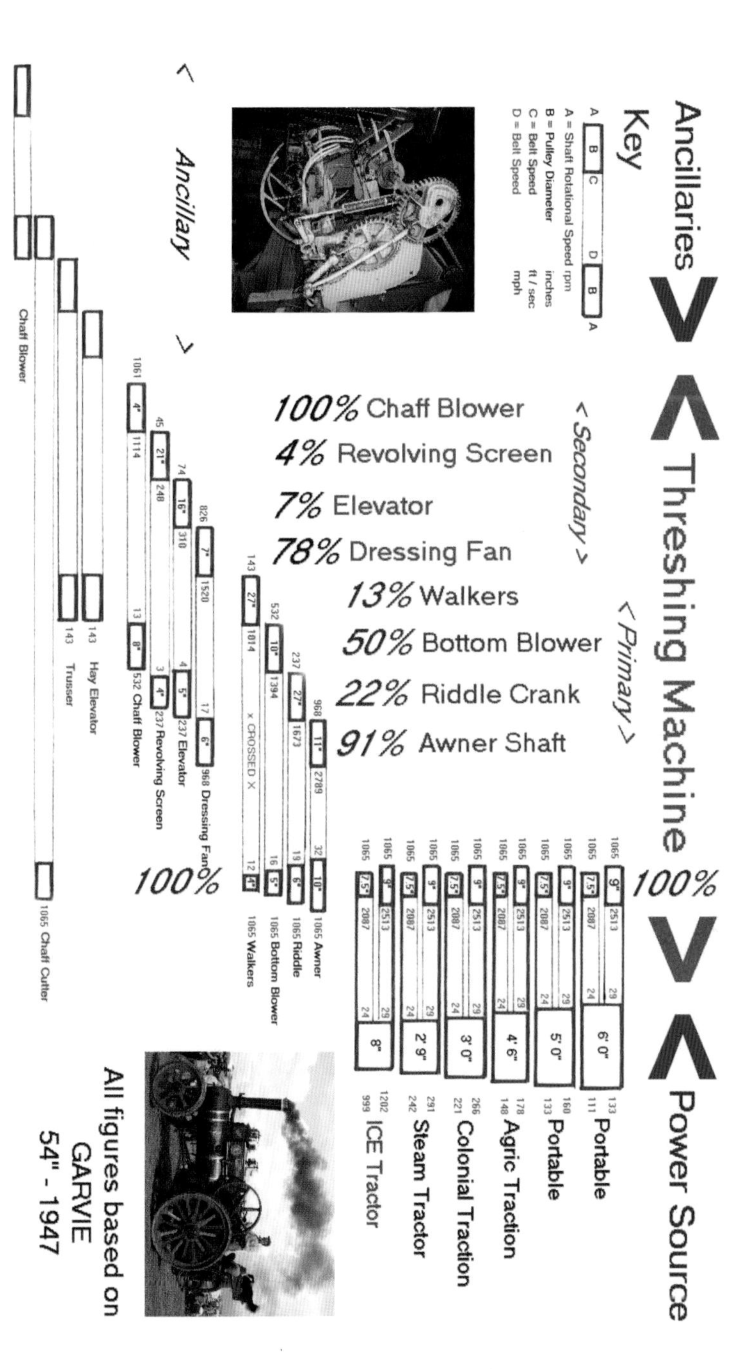

Chapter Eight

ADJUSTMENTS

Ransomes Instruction Plate

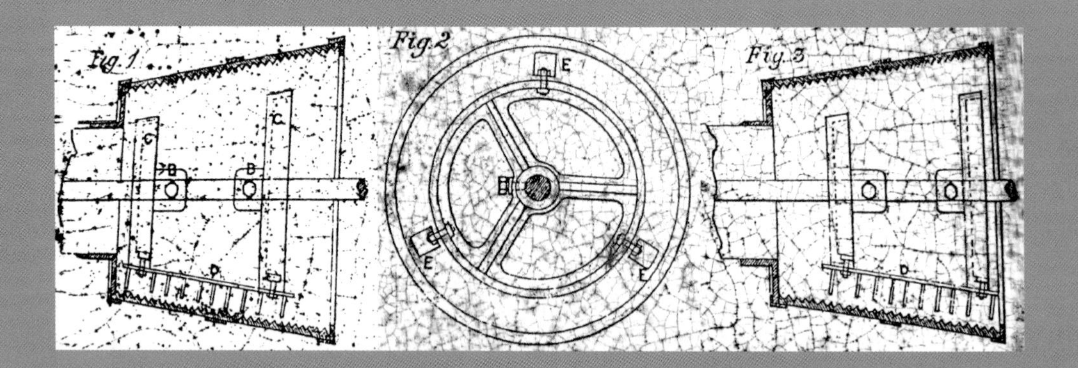

One of the great strengths of the threshing machine is its ability to handle the many variations in the crop to be processed. This ability is made possible in no small part through the great range of adjustments built into the design of so many parts of the machine, and it is for this reason that a whole chapter is devoted to this subject.

Firstly, in terms of sheer size of "grain", it can process larger items such as beans and peas. At the other extreme, it can deal with small seeds; it can even be adapted to hull clover. Indeed the purpose-built clover huller is based very much on the design of the thresher. In between, in its mainstream range so to speak, it can cover the full range of cereals; it can even cope with the mixture of cereals grown for animal fodder known as dredge corn.

Secondly, it is able to deal with crops infested with weed material. An 1881 article makes specific mention of processing crops abroad where the amount of weed material actually exceeds that of the crop itself.

Thirdly, it is expected to deal with enormous variations in moisture content in the material to be processed. At one extreme, a crop freshly cut in the rain; wet and clogging, in the majority of its work, fairly dry material from a rick, and at the other extreme, a crop freshly cut in tropical conditions, dry and brittle.

Fourthly, it can deliver a number of separate outputs and in a range of forms. Thus grain can be simply threshed and delivered as feed for livestock. It can also be cleaned and left with just a small number of awns in place to aid fermentation making it a perfect malting sample. Adjust the cleaning rate and a first class clean sample for flour milling can be produced. Not only is the grain delivered separately, but so are all the other outputs. The most important of these, straw, can be handled so carefully that it can be used for thatching. A development of this is the adaptation of the thresher known as a reed comber, where the stalk material becomes the primary output. Chaff can be fed to livestock. Weeds and seeds can be recycled as feed for poultry, or alternatively they can be burnt.

Although many mechanical adjustments could be made, by far the most important adjustment was that of the feeding rate. This required a great deal of skill on the part of the operators and this went hand-in-hand with the operation of the binder when cutting the crop initially. A combine harvester by definition has to handle the crop as a single operation and has far fewer options when poorer quality crops or conditions are encountered. This was very apparent in Britain when the early combines tried to replicate the performances they were able to achieve in the ideally dry conditions of California or Australia. As late as 1963, in his book "Farm Machinery", Claude Culpin states that it is not always possible for a combine to produce a perfect sample and that this may require secondary processing using barn machinery. The thresher, on the other hand, is expected to produce a perfect sample every time.

Beans and Peas

When shelling beans, the feeding safety cover was removed and the hinged top half of the concave would be moved back as far as possible. A shaped metal blank panel could then be inserted to cover the lower half of the concave. This was variously referred to as a bean plate, breast plate, or stripper plate. The middle and bottom concave adjustments were set to give a generous clearance, around ½" at the bottom. The cereal sieves would be replaced by models with the largest holes available and the second dressing part of the machine bypassed completely. For all cereal crops, the running speed was not generally varied, but in the case of bean shelling, the machine could be operated at around two thirds of normal speed, say 700 rpm. If much of this work was to be carried out, different sized pulleys could be fitted to allow the other moving parts to operate at normal speed whilst allowing the drum to rotate more slowly, but this was not common. Often a tent was erected over the drum and platform to contain the flying beans.

Seeds and Clover Hulling

Threshing machines could be used for obtaining seed from various crops. This was not possible with smallest of seeds such as Timothy but otherwise, this work was occasionally carried out. As in the case of bean shelling, the machine would be operated at a lower speed, generally around 700 rpm. Manufacturers could supply a blank metal plate to cover the back of the upper half of the concave. This was to prevent the heads slipping between the wires of the concave before the seeds could be threshed out. If such a plate was not available, it was possible to stuff the back of the concave with straw with a view to achieving the same result. The recommended clearances for the concave, top, middle, and bottom, were 1½", ¾", and 3/8" respectively. Manufacturers could also supply a special cavings riddle for seed work. This was in wood, like older thresher sieves. The first 2' were plain and from there on, the holes were drilled 3/8" or ½" diameter finished with a countersunk bit from below to prevent them from clogging. The first blast had to be kept fairly gentle, either by partially closing the shutters or, the better option of changing the belt pulley for a larger diameter item, thus reducing actual fan speed. The chob sieve of the main shoe would be a unit with 3/16" holes, and the dust sieve would be replaced with a blank plate on the basis that, in this mode, the material required was the smallest size produced. The dressing shoe top sieve would have 1/8" diameter holes and the bottom sieve 3/32". Here again, the blast from the dressing fan had to be gentle. The awner and chobber could be used, especially if the seed spikelets needed breaking up. The settings for the piler had to be very carefully chosen: too gentle and the unwanted material would not be removed; too violent and the seed could be skinned by too vigorous an action. The rotary screen was closed up to its tightest setting and the best seed was collected from the first spout which would normally have been for dust. The larger material retained by the screen that came off at the other spouts would be put through the machine for a second time. Since the total amount of material produced was very much less than that with conventional threshing, the loss of the use of changeover slides on the best corn spouts did not present a problem.

Clover Hullers

Purpose-built machines known as clover hullers were produced by several threshing machine builders. The best-known was the Ruston-Hornsby the production of which was later taken over by Ransomes after a company merger. Some conventional threshing machines could be adapted specifically for clover hulling and it is possible to hull the pin-head sized seeds from clover using a conventional machine. In order to do so, the material needs to be passed through the machine twice, first time to free the seed from the stalks and second time to rub the seed clear from the chaff. This is much less productive than conventional threshing and a full day's work can result in only 5 sacks of seed.

Drum

The drum itself is fitted with a set of replaceable beater bars. These are available in a range of patterns as described elsewhere. For certain larger crops, a better separation can be achieved when the drum is run with alternate bars removed. Because of the very high rotational speed involved, the drum will require re-balancing when these bars are refitted. Many beater bars are designed so that they can be reversed or rotated so as to present a new edge after wear has taken place. Wear on the bars also tends to be greater at one end because those feeding the machine will be naturally right or left handed and the heads will always be touched by the same piece of bar. Many manufacturers therefore designed the bars in such a way that they could be turned end-for-end to equalise this wear.

Concave

The Concave covers roughly half of the circumference of drum, this being known as the "wrap angle". It is constructed of square-edged bars usually spaced around 6" apart with 3/8" diameter wire threaded through holes spaced at around 3/8" centres. On most machines, the concave is in two halves hinged in the middle, and some of these use different wire spacings for the two halves. On only the smallest of machines is the concave in one piece. Whilst the clearance from the drum is adjustable, the spacing of the wires is not. Drum clearance is adjusted by means of pairs of eye bolts through which pass the rods holding the concave. Generally the eye bolts are set two horizontally and one vertically, and these pass through brackets on the outside of the cast-iron drum bearing casing. Once adjusted, the eye bolts are held in the desired position by a pair of lock nuts. The smallest machine with its one-piece concave will have only two pairs of adjusters in lighter brackets but the principle remains the same.

Fig 160: Concave adjustment (Ransomes)

The usual clearance settings can best be described as "double and double again". For a given gap at the bottom of the drum, this is doubled at the concave centre joint and doubled yet again at the top. Thus for ordinary wheat and barley, the clearances could be ¼" or less at the bottom, ½" at the middle, and 1" at the top. Oats, rye and lighter wheat would require the clearances to be reduced but these would still remain roughly the same proportion. Much of this adjustment is aimed at the careful treatment of the straw, the drier and more brittle, the greater the clearance. So as a season progressed, the clearance would be increased. Optimum setting lies between too wide when grain is seen leaving with the straw, and too narrow when the grain becomes bruised or broken.

The concave clearances are checked visually through a number of spy-holes in the main bearing casing at the ends of the concave. These holes are covered by swinging escutcheons similar in pattern to those that protect the two spirit levels. If threshed barley heads appeared in the cavings, one popular adjustment was to place Straw behind the top half of the concave. This was known as "stuffing the breast" and treated the crop more severely. A better alternative was thought to be to bypass the awner, of which more later.

Shakers

The shakers are designed to take the straw away from the drum and to separate the remainder of the grain that has not fallen through the concave but remains mixed with the straw. There are a number of small variations in shaker design but none are designed to offer a facility for adjustment as such. Some were modified by their owners by the fitting of a number of projecting coach bolts to improve flow. The speed of the passage of the straw is controlled by sets of doors or shutters that are suspended over the top of the shakers. These are simply hinged at their top ends and fitted with short check chains, fixed at one end and slipped over a hook at the other. The passage of the straw can thus be slowed by lengthening the chains which lower the shutters. Optimum setting can be said to have been achieved when movement of the straw is slowed sufficiently to separate the maximum amount of grain without the machine blocking up. The technical term for this is to obtain a thin and even "mat".

Shoes

The two shoes are the reciprocating frames that holds the various riddles and sieves. It is the bottom shoe that carries out the primary separation. The top shoe carries out the second separation or dressing and this is usually attached to the grain receiving board or tray below the shakers and the concave that moves the grain to the centre of the machine, where it can fall into the bottom shoe. On some smaller machines, the top shoe extends beyond the end of the shakers to help carry the straw away from the machine. Both shoes are given reciprocating motion from the riddle or jog crank. There is no facility for adjusting their speed or travel.

Riddles and Sieves

Very often the use of these two terms is completely interchangeable. When any differentiation is made, the riddles will be of a coarser type with larger holes and the sieves will be of a finer pattern. On modern machines, all are made from flat metal sheet, set in a timber sub-frame, and perforated all over with round or oval holes, all of the same size and evenly spaced. Each sheet is mounted at an angle of about 10 degrees.

The first sieve known as the cavings riddle is made up of a number of these sheets arranged in a gently stepped pattern; typically there would be 5 or 6 but fewer on some earlier machines.

Below this, the bottom shoe contains further sieves. The number and layout of these sieves varies with different manufacturers, each using their own preferred system. However the principle remains the same. The hole size of each sieve is slightly smaller than the one above gradually sifting out smaller and smaller material. Only the last sieve retains the grain allowing the remaining unwanted material to pass through. All but the last pair of sieves are fixed and are not intended to be adjusted. One of these upper sieves will be the chaff sieve and the first blast from the main fan will be directed towards its underside. The bottom pair, unlike all those above, are specifically designed to be quickly and easily changed according to grain size.

The smaller dressing shoe contains just 2 sieves and these are of the same quick-change pattern as the two bottom sieves in the bottom shoe. Here the second blast from the dressing fan will be directed towards the underside of the upper sieve.

Typical Settings

Table 4: Concave Clearance Settings

Concave - clearances:		Top	Middle	Bottom	
Barley / Wheat / Oats		1½"	5/8" - ¾"	3/8"	
		Set closer when wet			
Sieves - Hole sizes:					
		Barley	Wheat	Oats	
Bottom Shoe	Top Sieve	5/16" - 3/8"	7/32"	½"	Chobs sifted off.
	Bottom Sieve		¼" - 5/16"		Small seeds pass
Dressing Shoe	Top Sieve		5/16" 5/8"		
	Bottom Sieve		¼"		

Fans

The fans provide the blasts for the winnowing action that removes larger but lighter material. The blasts are numbered up to 4, and get their name from the order in which they were incorporated in thresher design. The speed of the fans is not adjustable but the power of the blast can be varied by opening or closing the shutters at the ends of the fan casings. The shutters can be slid to the desired position and then locked in place. Typically in the closed position, about one third of the end of the fan casing is exposed. Fully open, this increases to around 90%, giving an adjustment of a factor of 3. Where the Third and fourth blasts are ducted from the main fan, the so-called divided blast system, these are adjusted independently from the main and dressing fans by means of a pivoted flap able to cover part of the slot at the end of the duct. Generally speaking, the combination of a stronger blast with a coarser pattern sieve is thought to be preferable to a gentler blast and a finer sieve. Too much blast with light grains like oats causes them to be blown back into the machine for reprocessing, something which can lead to blockages in the grain elevator.

The Dressing System

The passage of the grain through the dressing system is controlled by a set of metal shutters. Most of these can simply be operated by external handles. These shutters or slides are so arranged so that none, some, or all of the items of the second dressing apparatus can be used. A typical layout would work as follows:

Opening the first shutter would bypass the system completely, allowing the grain to fall directly to the best corn spouts where it could be bagged.

Closing the first shutter would take the grain into the end of the Awner.

Opening the second shutter would mean that the grain was treated only by the first part of the awner.

Closing the second shutter would mean that the grain was treated by the full length of the awner.

Closing the third shutter would mean that the grain was directed to the smutter. The end of the smutter is fitted with a pivoted shutter that can be adjusted rather like a weir overflow and then locked into position. The higher the overflow level, the longer the grain is treated in the piler. The grain falling from all but the first shutter is directed to the dressing shoe.

Piler

Fig 161: Sliding Plain pattern Smutter Blades in ribbed interior (Foster)

Most pilers are of a similar pattern. Inside a pair of fixed casings, rotates a shaft to which are attached a number of blades. Very rarely are the blades in the cylindrical awner chamber adjustable. Those in the smutter can be slid along the shaft via the slotted holes in the brackets. This alters the clearance between the blade and the inner surface of conical chamber.

Adjustments to the piler take two different forms; one is flaps & shutters, the other is knife settings. The flaps allow the grain to drop out of the piler at a choice of points along its length, the further along the instrument the grain is allowed to pass, the greater the treatment. The shutters are arranged to control the speed of the passage of the grain in each chamber, the longer the grain is held in each chamber, the more severe the treatment.

The method of attachment allows their angle on the shaft to be readily altered. A number of the blades can be removed, making sure that the rotating unit remains in balance. The blades are also available in a number of different patterns so that the whole set can be changed if so desired. All adjustments can be made through the openings in the casing.

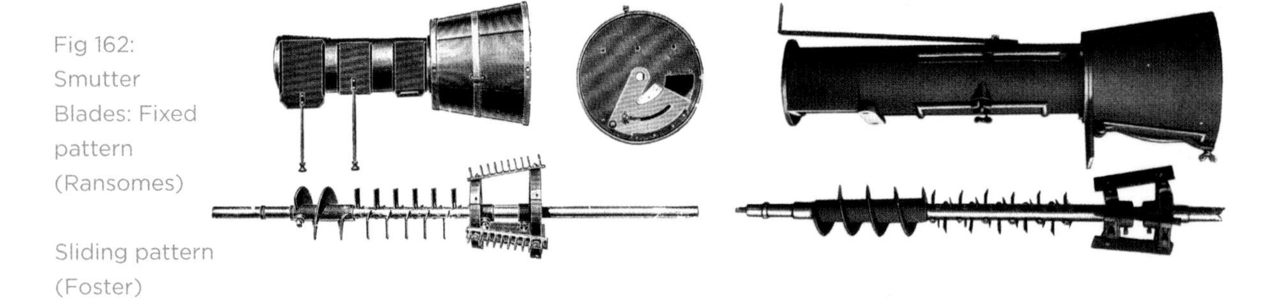

Fig 162: Smutter Blades: Fixed pattern (Ransomes)

Sliding pattern (Foster)

Rotary Screen

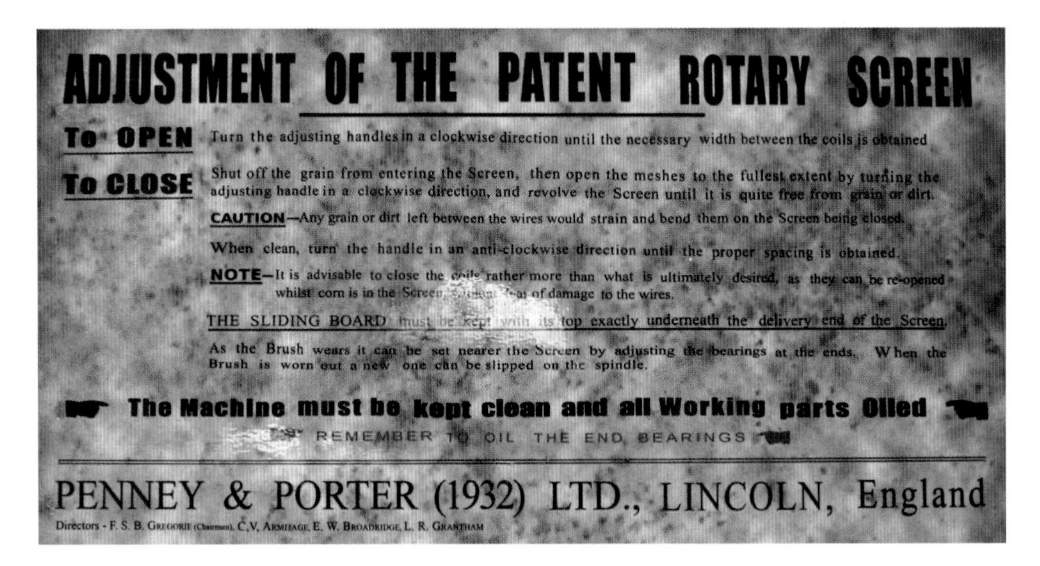

ADJUSTMENT OF THE PATENT ROTARY SCREEN

To OPEN Turn the adjusting handles in a clockwise direction until the necessary width between the coils is obtained

To CLOSE Shut off the grain from entering the Screen, then open the meshes to the fullest extent by turning the adjusting handle in a clockwise direction, and revolve the Screen until it is quite free from grain or dirt.

CAUTION—Any grain or dirt left between the wires would strain and bend them on the Screen being closed.

When clean, turn the handle in an anti-clockwise direction until the proper spacing is obtained.

NOTE—It is advisable to close the coils rather more than what is ultimately desired, as they can be re-opened whilst corn is in the Screen, [illegible] of damage to the wires.

THE SLIDING BOARD must be kept with its top exactly underneath the delivery end of the Screen.

As the Brush wears it can be set nearer the Screen by adjusting the bearings at the ends. When the Brush is worn out a new one can be slipped on the spindle.

The Machine must be kept clean and all Working parts Oiled

REMEMBER TO OIL THE END BEARINGS

PENNEY & PORTER (1932) LTD., LINCOLN, England

Directors - F. S. B. Gregorie (Chairman), C.V. Armitage, E. W. Broadridge, L. R. Grantham

Fig 163: Rotary Screen Instructions (Penney & Porter)

The grain from the dressing shoe is directed to a short auger feeding the rotary screen. This long cylinder of metal wire is adjusted by means of a handle with a square-section drive that fits into the fixed end of the screen. Turning the handle lengthens and shortens the screen which alters the gaps between the coils of wire. The cylinder is so designed that these gaps gradually widen away from the auger feed. Below the cylinder are a series of hoppers that lead to the corn spouts. These are so arranged that when the screen is correctly adjusted, the grain is sorted as follows:

The smallest material, dust etc, is directed to the first hopper.
Smaller grains or tailings are directed to the next hopper(s).
Best corn is directed to the last hopper which is double-sized.

The wires are kept clean by a rotating brush. The screen is easily opened, but if it has to be closed up, care has to be taken to ensure that no grains remain trapped between the wires as failure to do so can result in bent or broken wires. Before closing the screen, it has to be opened up fully and all loose material removed. This can be a lengthy operation as each grain has to be picked out individually.

Threshing in the West

Threshing Anecdotes

Local Village Produce Shows often featured a class for corn.
Each sample was presented in a plain hessian sack, not quite full, and with the top neatly folded back. A darker colour for the sack was to be preferred as this made the corn appear even more golden. The small size of the communities involved meant that everyone knew everyone else...

And more tellingly, whose produce was whose...

Year upon year, exhibitors and judges alike took up their respective roles, and year upon year, the prizes were wont to go in the same direction...

At one particular show, one particular year, a disgruntled corn exhibitor carefully buried a small selection of his cat's litter tray just a few inches below the surface of the grain. He knew that the regular, if somewhat unsympathetic, judge would perform the same ritual that he had performed so many times before. He would dive his hands into the sack, take out a sample, rub it firmly between his palms, and then hold it close to his nose. Of course, the judge would be unwilling to complain. To have done so would have been to admit to knowing the identity of the exhibitor...

Chapter Nine

SURVEYING THE MAKERS

The majority of threshing machine manufacturers at some time also produced the steam engines to drive them and could therefore offer complete "threshing sets". This was common earlier on with even smaller manufacturers turning out portable engines, these being far simpler to build than tractions. Thus the larger thresher producers built tractions, and the larger traction producers built threshers. The two biggest exceptions were Fowlers with no threshers, and Humphries with no tractions, although they had produced a few portables.

In terms of production, the largest number of threshers was produced by Marshalls, followed by Clayton & Shuttleworth. Large numbers were also produced by Fosters, Garretts, Humphreys, Ransomes and Rustons. Significant numbers were also produced by Barrow & Stewart, Burrell, Robey and Wallis & Steevens. Machines were also built in Scotland by no less than 17 firms including Allen, Barclay Ross & Tough, James Crichton, RG Garvie & Sons and AC Tullos to name but a few. None of these built steam engines.

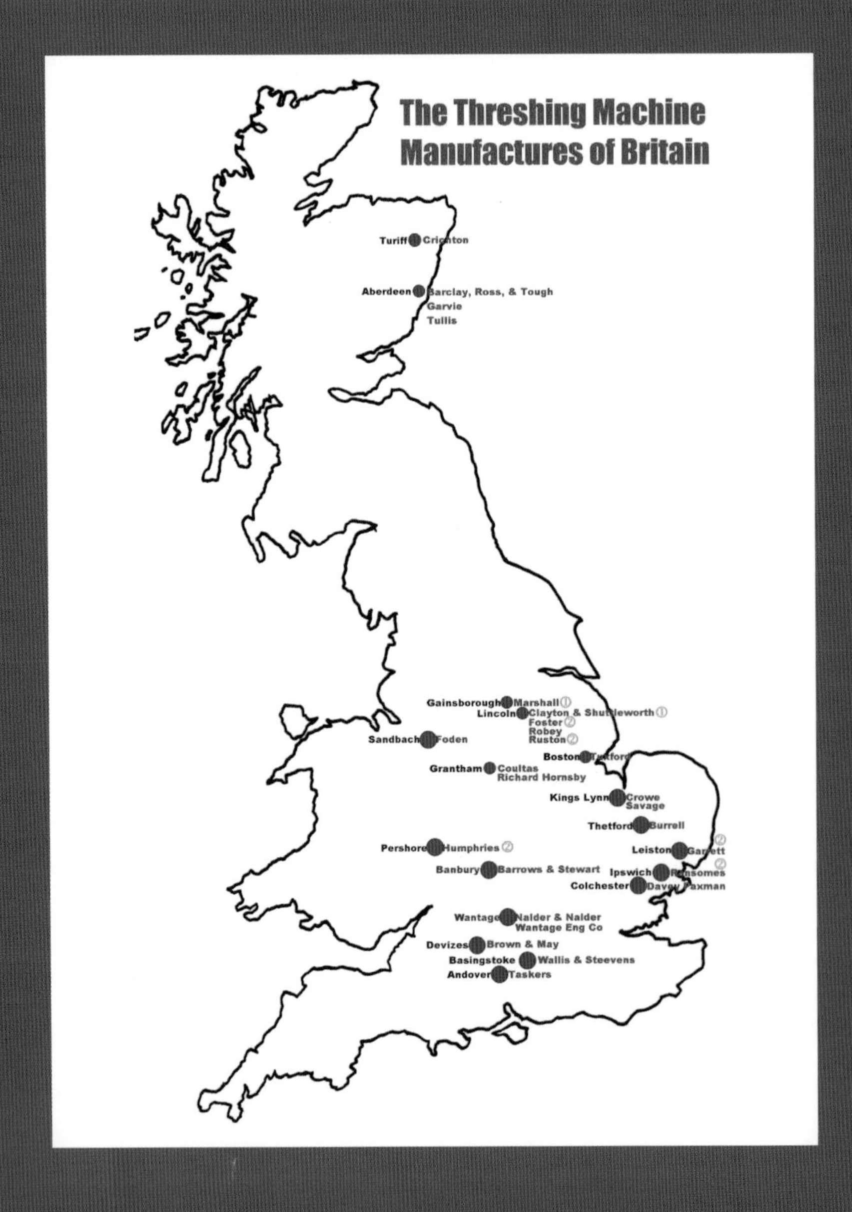

Fig 164: Map showing the Threshing Machine Manufacturers of Britain

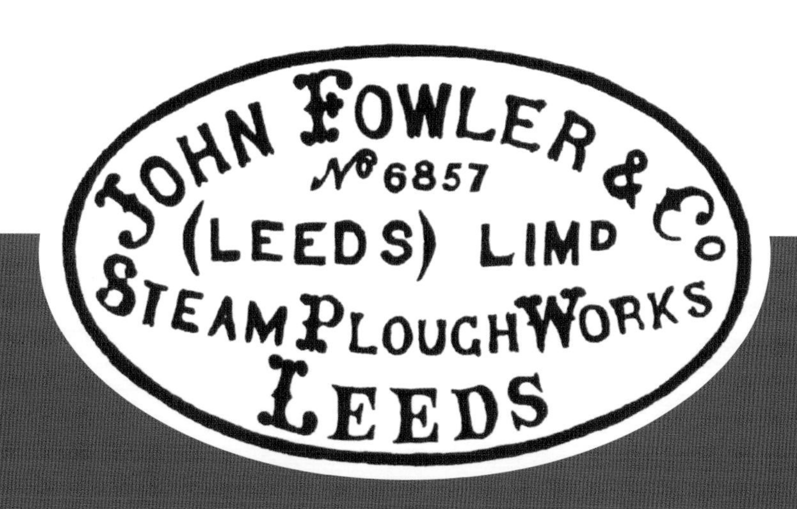

Aveling
Burrell
Clayton & Shuttleworth
Foden
Foster
Garrett
Marshall
Ransomes
Robey
Ruston
Savage
Tasker
Wallis & Steevens
Barrett, Exall & Andrews
Barrows
Davey-Paxman
Gibbons
Humphries
Tuxford

THOMAS AVELING

Early advertisements show that the Aveling company, in common with the majority of the manufacturers of portable steam engines, and later traction engines, also offered threshing machines. These were often sold together as a "threshing set". If Thomas Aveling was not a great innovator in the design of the thresher itself, he is more than worthy of inclusion for his pioneering work in areas closely associated with the threshing operation as a whole.

He was born in Cambridgeshire in 1824 but as a child, moved to Rochester, Kent, where he later became apprenticed to a farmer. He married his niece and began farming on his own account near Romney Marsh, but his interest in all things mechanical led him to open a small engineering works back in Rochester in 1850. An early design was for a system for ploughing by steam that was first demonstrated in 1856 and for which he received the then huge prize of 300 gns from a group of Kentish farmers two years later.

However it is for his key input into traction engine design that he is best remembered. In 1858, he converted a Clayton & Shuttleworth portable engine to make it self-propelling by driving a single rear wheel by pitch chain, a crude predecessor of the modern renold chain. Portables at the time were moved from place to place using teams of horses, and when made self-propelling, many used a single horse in shafts for steerage. This was thought to have the advantage that a horse in front of the machine was likely to reduce the risk of other horses shying when they saw the portable. Aveling dispensed with the horse completely for the first time using his patent fifth wheel pilot steerage. In 1861 he patented the system of steam-jacketed cylinders which was adopted by all other manufacturers, as was the extension of the firebox wrapper to produce hornplates to support all the gearing in 1870.

He took out a patent for a road locomotive in 1859 which was built for him by Clayton & Shuttleworth, that he exhibited at the 1860 RASE Show, held that year at nearby Canterbury. Claytons also built a traction engine for Aveling in the same year, but by 1862, he had started building his own machines exhibiting the first of these at the Leeds Royal Show and later at the International Exhibition. Within 12 months, no fewer than 97 engines had been produced.

If Thomas Aveling did not invent the steam roller, he can certainly claim to have made it his own. A machine had been produced for use in Calcutta in 1863, and within 3 years, the first Aveling prototype was built. The firm of Aveling & Porter went on to build more steam rollers than all other manufacturers put together. The other steam engines they produced were generally well thought of, but the roller part of the business, particularly with its substantial export business, limited development and production of other machines.

The events of World War One and its immediate aftermath ruined the company financially and forced it to join the ill-fated AGE - Agricultural and Engineering Group - in 1919, along with 13 other similar companies including Burrell, Garrett, and Davey-Paxman. The new consortium struggled on until finally collapsing in the great slump of 1932. Soon after, the original company re-formed as Aveling-Barford and went on to continue production of road rollers. Most of these would have been diesel powered but the very last steam rollers to be made in the early 1950's, still carried the famous prancing horse emblem - and the kentish motto, "Invicta".

BY HER MAJESTY'S ROYAL LETTERS PATENT.

AVELING'S PATENT STEAM THRASHING TRAIN.

ROCHESTER,

To THOMAS AVELING, Dr.

AGRICULTURAL ENGINEER AND IRONFOUNDER.

PORTABLE STEAM ENGINES, THRASHING MACHINES, SAW BENCHES, PUMPS, &c., FOR HIRE.
STEAM ENGINES, BOILERS, AND MACHINERY REPAIRED.

Every Description of Agricultural Machinery supplied at Manufacturers' Prices

Fig 165: Aveling Steam Threshing Train

Fig 166: Aveling Chain-drive Engine

Fig 167: Aveling Insignia

Fig 168: Aveling Steam Roller

Fig 169: Aveling Showman's Engine

Chas Burrell & Sons

Fig 165:
Aveling Steam
Threshing Train

St Nicholas Works, Thetford, Norfolk

The traction engines produced by Charles Burrell & Sons are thought by many to represent the finest ever built. The firm exhibited its first portable in 1848 and this formed the basis for a number of self-moving machines using the wheels patented by James Boydell. Burrells continued using the Boydell patent from 1856 up to 1862, often on a purpose-built machine. Similarly in the next decade, they also built a number of rubber-tyred road steamers to the design of Robert William Thompson. Their early mainstream traction engines, like many of their competitors, used chain drive and with the cylinders over the firebox in portable style, thes remained in production for over 10 years until the late 1870's.

In search of improved efficiency, the company experimented with compounding. A tandem design was exhibited at the 1881 RASE Show held at Derby, and many conventional double-crank or "cross" compounds were produced by the company starting in 1896. However they are best remembered for their single-crank compound design that was the subject of patent number 3489 of 1889. Early models gained a reputation for uneven running nicknamed the "burrell thump" but this was cured on later models by the fitting of a balanced crank. This design formed the basis for one of their most popular of agricultural engines, the lightweight 6 nhp "Devonshire" class that was introduced in 1897 when it cost £475. At the other end of the scale, the famous showmans road locomotive, "Lord Roberts" appeared in the same year. The company went on to become the largest producer of this class of engine which culminated in the 10 nhp "Special Scenic" engines with their double dynamos to power the scenic showground rides with their exceptionally heavy starting loads. Their steam tractors were also very successful with the 5 ton version taking first prize and the gold medal at the 1907 RAC Trials.

As with many of their competitors, the company struggled after WW1, and joined the AGE consortium. Many of the engines built after the amalgamation carried the AGE logo on their smokebox doors. They produced their last showmans engine, "Simplicity" in 1930 but the consortium failed just 2 years later. A mark of the quality of the products and the affection that they engendered is the fact that of the 4094 engines produced, over 10% remain in preservation.

Although the company is best known for its traction engines, it made a major contribution to the Threshing industry with its machine of 1848 which is the first to offer the integral secondary dressing mechanism that defines the "modern" thresher.

Fig 170: Burrell's first "Combined" Thresher

Fig 171: Burrell Single-crank Compound

Fig 172: Burrell Threshing Set

Fig 173: Burrell "Special Scenic" Showman's Road Locomotive

Fig 174: Burrell late Insignia

Clayton & Shuttleworth

This company was founded by Nathaniel Clayton and Joseph Shuttleworth in 1842 as an engineering and iron-founding business making iron pipes. The business soon branched out into steam-powered farming machinery, and was one of the first to build portable engines. Their 1848 design machines featured a wood-lagged cylinder over the firebox with a crankshaft forged from the round in a pair of cast brackets towards the smoke box. These all featured a 12" stroke and the bore size varied according to the desired power output. Thus 4 nhp used a 6¼" bore, 5 nhp = 7" bore and 6 nhp = 7¾" bore. By 1852 the workforce had risen to 80, and within 4 years, the number of portables produced exceeded 2200. Their early entry into portable production meant that other manufacturers based their first designs on C & S practice, and not a few commissioned the company to build machines to their own designs.

A visitor to a modern steam rally could be forgiven for thinking that Victorian steam engine production consisted of many traction engines and a few portables. In fact the exact reverse is true. In the same vein, it might be thought that the portable was the forerunner of the traction and that its production ceased once the traction became established. Once again this most certainly was not the case. Portable engines were built alongside tractions and their production continued long after that of tractions, steam tractors and even steam lorries had ceased. To put this into context if 2500 portable were built between 1846 and 1856, then in 1913 alone, portable output was 1297.

By 1861, a workforce of 940 were turning out 15 machines a week and in 1870 alone, 1200 men built a total of 1800 portables. As the firm entered the 20th century, over 2000 men were working for a Limited Company whose sales literature proudly boasted "58 000 engines and machines sold". In 1900 the company had expanded its Eastern European export business by joining with Matyas Hofherr and Janos Schrantz of Budapest, Hungary, to form Hofherr-Schrantz-Clayton-Shuttleworth. During the First World War, the company built Handley-Page and Vickers bombers even converting part of the factory site to an airfield.

Unfortunately this financial success during the war could not prevent a collapse in 1919. The export drive into Eastern Europe before the war had resulted in many orders from Russia. After the 1917 revolution, the new Bolshevik government repudiated all foreign debt. As with their rivals, Claytons joined the ill-fated AGE group lasting until 1929 when they became part of Marshalls. In these later years they produced a range of steam vehicles including wagons as well as rollers and tractors.

Fig 175: Clayton & Shuttleworth late-pattern Thresher

They also to continued to build threshing machines that had always formed a major part of the company's output. Peak production was reached in 1913 with a total of 1211. The machines were well-built, well designed, and well thought of. They were the main proponents of single-crank shaker design, but they too adopted the double-crank layout after 1915 for the last machines. For a time, they used a blower arrangement instead of a grain elevator, but reverted to a conventional design. The expertise in this area of machinery persuaded them to build the first British combine harvester in 1928. For many reasons this was not a success and the combine did not achieve acceptance in Britain until the arrival of the Lend-Lease American machines imported during the Second World War.

Fig 176: Clayton & Shuttleworth early-pattern Portable

Fig 178: Clayton & Shuttleworth late-pattern Roller

Fig 179: Clayton & Shuttleworth 1896 Traction

Fig 177: Clayton & Shuttleworth nameplate

Fodens

Elworth Foundry, Sandbach, Cheshire

The name Foden is associated with steam wagons and more recently, with their diesel-engined successors. It might be a surprise to learn that the Foden company began with a separate section devoted to threshing contracting and that the early traction engines were regarded as some of the most efficient of the era.

Founder of the company, Edwin Foden was born in 1841 and in 1856, was apprenticed to George Hancock, son of the famous Walter Hancock whose steam road carriages dared to challenge the railway monopoly of the 1830's. The company of Hancock & Foden was formed in 1876 and their first traction engine appeared 4 years later. Despite their relatively late entry into the traction engine market, the company quickly established a reputation for quality and efficiency. Their engine with duplex high-pressure cylinders won a gold medal at the 1887 RASE Show held that year at Newcastle. Their performance was superior to many of their compound-engined rivals and not surprisingly, their own compound designs with their patent 3-way cocks giving a "double-high" facility set a standard that few could match.

In 1898, the company produced their first steam wagon which went on to establish a dominance of a market that few had foreseen. By 1934 when the last wagon was built, the company had produced no fewer than 7000 machines, very many more than were produced by all their rivals put together. The early wagons were of the "overtype" design and surprisingly, this went against the results of the government trials organised in the early years of the 20th century. Despite this success, Fodens changed to "undertype" wagons in the mid 1920's, and successfully challenged Sentinels who had previously dominated this market. At a time when other builders of steam engined vehicles were going to the wall. Fodens turned to the production of diesel-engined vehicles, an approach which they are able to continue up to the present day.

One might be excused for thinking that the above history has little to do with Threshing. In fact Fodens were a recognised builder of threshing machines and in the early years of the firm, Edwin's sons ran a separate branch of the company as threshing contractors. In later years, many of their D class steam tractors ended their working lives powering threshing sets. Due to the punitive legislation of the 1930's, some steam wagons were cut down as tractors and some of these would have spent their closing years like their siblings, threshing. All in all, an intriguing close to a circle of technical development.

Fig 180: Foden Thresher

Fig 181: Foden Smokebox Door

Fig 182: Foden early-pattern Traction

Fig 183: Foden D-type Tractor

Fosters

Wellington Foundry, Lincoln

William Foster was born in 1816 and started his business life as a flour miller. He founded the engineering company to bear his name at Potterhanworth, 6 miles southeast of Lincoln in 1856. Within 3 years, the flour milling business was wound up so that the company could concentrate on the manufacture of farm machinery.

Their portable engine No 64 was exhibited at the 1861 Smithfield Show and production of these and traction engines continued steadily up to the turn of the century. General improvements were made, particularly changeovers from wrought iron to mild steel for the boiler plates, and to cast steel for the gearing in the early 1870's. Never the largest of companies, the workforce grew from around 70 in 1871 to over 350 by 1914. Nearly 600 tractions were made from 1880 onwards, but Foster's heyday for steam engine production was later than that of many of its rivals. However, it also continued for longer. The first of 88 road locomotives was not built until 1904 but the last appeared in 1934. In fact the last traction engine built by any manufacturer was a Foster completed in 1938. The vast majority of the road locomotives were built as showmans engines and not a few of the others were later converted for the same purpose. The "second generation" of these were of the special scenic type with an integral crane and a second dynamo capable of delivering a massive 200 amps at 110 volts. During this period the company also produced a very successful range of steam tractors; 350 machines in 5 different marks with the 5 ton version taking the Silver Medal in the 1907 RAC Trials. This technology proved useful in the development of their overtype steam wagon. In 1909, Fosters built a steam caterpillar tractor to a design by the Richard Hornsby company, for use in the Yukon goldfields. This experience persuaded the government to choose the company to build the First World War tanks. After the war, the outline of a tank was incorporated into the nameplates on their castings, although this was dropped on the very last models. Fosters took over Gwynne Pumps in 1927 and became a major producer of land drainage machinery. As part of the takeover deal, they also supervised production of the "Albert" car company.

In parallel with their output of steam vehicles, Fosters were a major producer of threshing machines. They have been called by some, "The Rolls-Royce of Threshers". The first of these appeared in 1848, and a further 20 had been built by 1854. The first to feature shakers and a riddle was exhibited in 1852 and a number of accolades were won by derivatives of this design: Commendation at the 1858 RASE Chester Show, Special Mention at the 1858 Yorkshire Show held at Northallerton, and First at the 1859 Highland Show (held at Edinburgh). Between 1858 and 1889, no fewer than 1320 threshers were sold, many as threshing sets, which accounted for many of the 1400 portable engines sold over the same period. A substantial proportion of this output was destined for export, particularly to Central and South America in the 1890's. In total, over 500 threshing machines were built. Just as the last traction engine made was a Foster, so was the last threshing machine, sold to a farmer in Louth, Scotland, in 1961, just 5 years before the company folded.

Fig 184: Foster Thresher c 1900

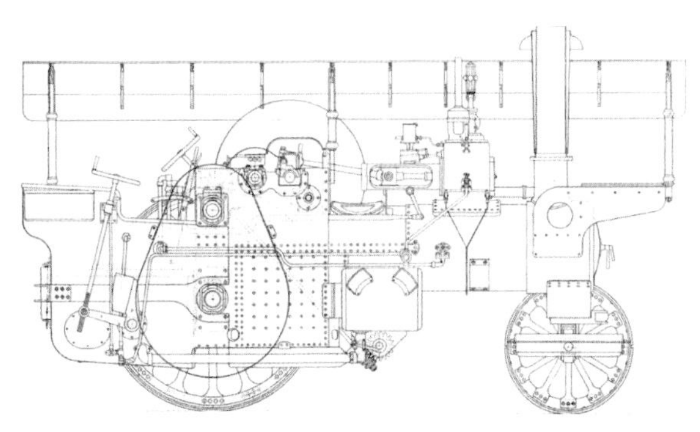

Fig 185: Foster late-pattern Portable

Fig 186: Foster early-pattern Showman's Engine

Fig 187: The last Thresher ever built (Foster 1961)

Fig 188: Foster late-pattern Showman's Engine

Richard Garrett

Leiston Works, Leiston, Saxmundham, Suffolk

The family firm of Garretts of Leiston has been traced back as far as 1674 and was responsible for many early innovations in the field of farm machinery. In1805 Richard Garrett married Sarah Balls, the daughter of John Balls who designed the first open concave. This eventually became adopted by every thresher manufacturer and the marriage must have given Garretts a great head start. This was followed in 1846 by perhaps the biggest single step forward in thresher design, the "bolting" thresher, where the crop could be fed crosswise resulting in an enormous reduction in damage to straw.

The earliest Garrett threshers were powered by horse-driven capstans of their own manufacture. There is a claim that a steam portable engine was built in 1824 but this appears to be in some doubt. What is known is that a Garrett portable was exhibited at the 1840 Norwich Show. Two Portables were exhibited at the 1848 York Royal Show one of which won the RASE first prize. These machines went on to win further prizes at various shows over the next few years and a number appeared at the Great Exhibition of 1851. The first self-propelling engine featuring Boydell's patent wheels was exhibited in 1856 and a number of other designs followed soon after, including a portable with pitch-chain drive to a single rear wheel with horse steerage, and a digger based on Cooper's patent. The first conventional traction engine was built in 1867 and production continued for the next 60 years. Compared with some other companies, traction engine output was not great, but the 514 steam tractors sold between 1904 and 1929 made Garretts' the most popular marque. In 1919, an attempt was made to compete directly with the internal-combustion tractors of the time in the form of the "Suffiolk Punch", an unconventional lightweight steam tractor designed by Denzil Lobley. It was not a commercial success, neither was a second version produced in 1930. By comparison, the last portable left the factory in 1940.

The size of the company grew steadily from 300 employees in 1851, doubling to 600 by 1862, and doubling again to 1200 in 1910. A major fire destroyed much of the factory premises in 1913, but the company succeeded in maintaining production, continuing to enjoy a thriving export market that included Southern as well as Eastern Europe. Unfortunately, the Russian revolution resulted in the repudiation of debts which, in Garretts' case, exceeded £200,000. In common with its rival firms faced with similar financial problems, Garretts joined with 13 other companies to form the AGE group, only to be taken over when that failed in 1932, by Beyer-Peacock. As an aside, unlike so many of their rivals, the company records and drawings still exist almost in their entirety. These give an excellent insight into the whole era, not just the technical and financial aspects of the company, but also the history of the area.

The firm was the 5th largest producer of threshing machines and can claim to have made more innovations in threshing design than any other company. After the wire concave and bolting drum, the firm specialised in the fitting of fluted beater bars, and in 1859, produced its famous single-fan machine. Beater bar design was improved by adopting a reversible pattern. An improved multi-fan design finally took over from the earlier single-fan threshers but these remained in production until as late as 1900. Most machines exported by British manufacturers were very much of the traditional "English" pattern but Garretts for one, and Marshalls for another, built some on the lines of American threshers with steel frames and cladding, apron style shakers and even "wind stackers".

Fig 189: Garrett Thresher c 1900

Fig 190: Garrett export-pattern Portable

Fig 191: Garrett 4CD Steam Tractor

Fig 192: Garrett "Suffolk Punch" MK 1

Fig 193: Garrett superheater Road Locomotive

Marshall

William Marshall (1812-1861) founded his company of "Millwrights and Engineers in 1842, with the Britannia Iron Works opening just 5 years later. Their first portable was completed in 1857, and the first traction engine in 1876, in both cases somewhat later than those of Marshalls' main rivals. The traction was an undertype thus avoiding infringement of Aveling's hornplate patent. Unfortunately it suffered from the disadvantages that it was heavy with a high centre of gravity, and also that the inaccessible flywheel made attachment of a belt extremely difficult. A traction engine of conventional layout appeared in the following year and the first road locomotive in 1883. The company expanded into one of the largest producers with the workforce of 600 in 1870 doubling to 1200 in just 6 years peaking in1917 at no fewer than 6000. Sales literature quoted sales of engines and boilers exceeding 75,000 before 1900 rising to 189,000 by 1925. A substantial proportion of this output was built for export and a range of "Colonial" tractions, many designed as straw burners proved particularly popular.

Unfortunately the outcome of the First World War resulted in non-payment of heavy overseas debts forcing the company, like so many others, to join the AGE consortium. Marshalls were able to take over the orders for threshers from their main rivals, Clayton & Shuttleworth and somehow managed to avoid the crash of the AGE in 1932. As far back as 1881, the firm had obtained a licence to manufacture the Otto gas engine, and this early involvement may have helped them with the development of their famous range of single-cylinder diesel Tractors in the 1930's. Production of "Field Marshalls" continued into the 1950's, to be joined by a number of Combine Harvester designs.

Marshalls were Britain's biggest builder of threshing machines receiving a Show prize as early as 1849. Gold medals for threshing sets were won at the Paris Universal Exhibition in 1868, and the RASE thresher trials held at Cardiff in 1872. A huge variety of machines was produced with drums ranging from 3' 0" to 5' 6" wide in 6" increments, each available in medium or heavy duty form, and each in a number of forms offering varying degrees of finishing quality. Marshalls were one of only a few companies to offer the option of a fourth blast. Unlike their rivals, this was ducted from a separate third fan mounted on the extended end of the second dressing fan shaft, and set to blow above the rotary screen.

Interestingly, the building of machines framed in timber ceased in 1931, to be replaced entirely by steel. This was despite the fact that in the 1920's, 3 out of 4 threshers sold were timber-framed. The early steel-framed machines featured a bottom shoe with a shorter travel and running at a higher speed than most other machines which proved to be a less efficient design. In 1929, this was corrected by reverting to the older layout for the crankshaft and larger pulleys. Marshalls acquiring the manufacturing rights for Clayton and Shuttleworth products enabled them eventually to incorporate their better awner design.

At the start of World War Two, the government was anxious that more food should be grown at home. Farmers when ordering new threshers demanded their traditional timber-framed machines and Marshalls were forced to reverse their build policy. Nonetheless, over 200 threshers were built between 1939 an 1945, more than the combined output of all other manufacturers. These were all supplied with a steel forecarriage based on the design fitted to their steel framed machines. Like Garretts, they also built a small number of machines especially for export in the American style as well as the conventional "English" pattern. Following on from C & S's attempts in the late 1920's, Marshalls brought out a range of trailed combine harvesters immediately after WW2.

Fig 194: Marshall Steel-frame Thresher c 1930

Fig 195: Marshall large Portable

Fig 196: Marshall late-pattern Traction Engine

Fig 197: Field-Marshall tractor

Fig: 198 Marshall undertype Traction

Ransomes, Sims, & Jefferies

Also Known as: Ransome & May, Ransomes, Sims, & Head

Ransome's were one of the older companies to be in involved in the threshing business. Robert Ransome (1753-1830) opened a small iron foundry which went on to produce a whole range of goods particularly agricultural equipment. In 1785, he took out a patent for hardening cast-iron ploughshares, the so-called chilled iron tempering process. This was followed in 1808 by another patent, this time for interchangeable plough parts. Away from agriculture, the firm also supplied and installed Norwich's first gasometer in 1817 and built a cast-iron bridge in Stoke-on-Trent. In 1837, their famous Orwell Works opened and 2 years later, the first of a new line of machinery was built, a product that was to become synonymous with the firm, Lawn Mowers.

The firm built a number of portable engines, the first of these being exhibited at the 1841 Liverpool Royal Show. The following year, the first self-moving portable was exhibited at the Bristol Royal Show. This was driven by a pitch chain to a single rear wheel and steered by a horse in shafts. It carried a vertical boiler supplying steam to a rotary engine built to Davis' patent. More interestingly from the point of view of this book, the chassis of the vehicle was extended to carry a small threshing machine. The mobile threshing set had been born. In 1849, Ransomes received first prize at the 1849 Leeds Royal Show when they exhibited their "Farmers Engine". This was designed by Robert Willis and built by E B Wilson, Railway Foundry, Leeds. It was a duplex cylindered undertype designed to be jacked up underneath the firebox so that other stationary equipment could be driven off the rear wheels, either by belt or via universal jointed shafts. The company also built one of Thompson's patent road steamers in 1871, before concentrating on their own range of more conventional Tractions. They also built a number of steam tractors beginning in 1903. After the Second World War, the company joined with Ford to build its Fordson-Ransomes implement range.

The company was one of the major producers of threshing machines, and the existing company records have allowed much of the detail of the early machines to still be available for study. The last quarter of the 19th century saw the heyday of thresher production when 80% of these machines were exported. After 1900, the market shrank drastically. The revolution meant that Russia was lost as a customer, the home market was very limited, and the New World was dominated by American machines. Despite this, the firm built its last thresher as late as 1954.

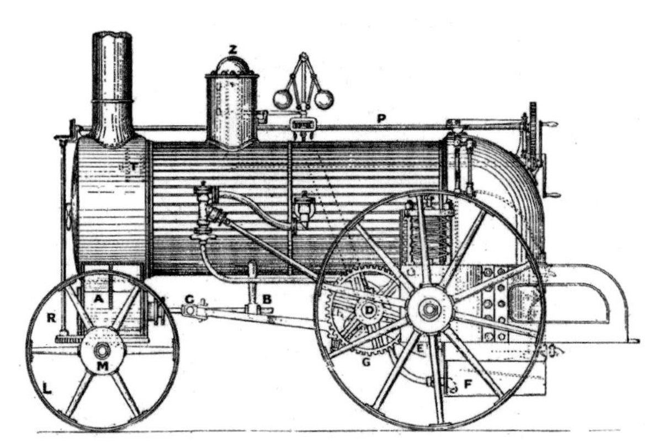

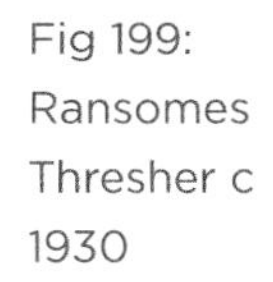

Fig 199: Ransomes Thresher c 1930

Fig 200: Ransomes "Farmers Engine"

Fig 201: Ransomes Steam Tractor

Fig 202: Ransomes late-pattern Portable

Fig 203: Ransomes Thresher c1880

Robey

Globe Works, Lincoln

Lincoln was a major centre for the building of agricultural steam engines and threshers. Of the 4 best-known makers, Robey's were the smallest. Founded in 1854 by Robert Robey, they had built a good number of conventional portable engines before exhibiting their first self-propelling model at the 1861 Smithfield Show, which carried the works number 938. In the same year, they built a ploughing engine to the design of W Savory of Gloucester with the winding drum arranged to rotate around the boiler shell. In common with a number of their rivals, Robeys built a Thompson Road Steamer in 1870 and a curious jackshaft traction to Box's patent. In later years they went on to build road rollers, including some to tandem pattern designed for use with the new soft tarmac materials. They built a number of steam wagons and one of their most interesting designs was their High-speed tractor of 1927 designed with pneumatic tyres, to compete with the Foden D Type and the Mk 2 Garrett Suffolk Punch.

Their threshing machines were not as numerous as some but showed an individuality in design as did their other products. They used a unique form of single-crank shaker mechanism and some where fitted with a fixed grain board with a conveyor-type action rather in the style of American "apron" machines.

Fig 204: Robey early-pattern Thresher

Fig 205: Robey early-pattern Traction

Fig 206: Robey late-pattern round-firebox Portable

Fig 207: Robey "High-speed" Tractor

Fig 208: Robey Tandem Roller

Rustons

Also known as: Proctor & Burton, Burton & Proctor, Ruston Proctor, Richard Hornsby

Spittlegate Works, Grantham, Lincolnshire Sheaf Iron Works, Lincoln

This group of businesses, all involved with steam engine and thresher production seem to have changed the format of their company more than any other. Richard Hornsby (1790-1864), founded his own company in 1815. He won a £50.00 prize at the 1848 York Royal Show with a portable featuring a round stayless firebox, a design that was emulated by a number of other manufacturers. Hornsby Portables won RASE first prizes in 1851 and 1852, and again in 1858 at the Royal Chester Show. An undertype traction engine appeared in 1863 to patent no 1726 registered in the names of Hornsby, Bonnall, and Astbury. A conventional design traction did not arrive until 1876. The business became a limited company in 1880, and the last steam engine to carry the Hornsby name was built in 1904. Seven years before that, a Hornsby tractor won a Silver Medal at the Manchester Royal Show powered by a Stuart & Binney hot-bulb diesel engine and in 1909, produced a tracklaying vehicle with an internal combustion engine.

Joseph Ruston (1835 to 1897) established the firm of Burton & Proctor in 1850. This changed its name to Ruston, Proctor & Co in 1857. They exhibited a portable at the 1861 Smithfield Show fitted with a new invention, the Giffard steam injector. During the First World War, Ruston, Proctor & Co Ltd merged with Richard Hornsby to form Ruston & Hornsby Ltd. They built their last traction engine in 1936, before finally amalgamating with Davey-Paxman of Colchester in 1940.

Some early threshing machines were arranged so that the cavings riddle moved with bottom shoe as one unit driven off single throws of the crank. This arrangement proved very prone to blockage and later designs reverted to the normal pattern of the two units operating 180 degrees apart. Another variation was to drive both shoes off additional throws on the shaker crankshaft. A similar arrangement was used on the Wantage Engineering Company's "Simplissima" machines. Both suffered from the disadvantage that all the drives for reciprocating parts were taken from a single small pulley which caused considerable problems of slippage unless the belt tension was adjusted perfectly. Machines built after 1870 were constructed to the more common pattern. Company records show that in the 40 years up to 1897, the companies produced a total of 20,800 steam engines and 10,900 threshing machines. When they were part of the Agricultural and General Engineering Group, production of their popular patent clover huller was taken over by Ransomes, another member of the same group at the time.

A testimonial letter in a catalogue from 1907 is typical of many of the period. "The Portable Engine no 714 delivered in to us in 1865, is still at work and will last for many years to come, giving continual good work".

Fig 209: Early Ruston Proctor Threshing Set

Fig 210: Late-pattern Ruston Portable

Fig 211: Ruston Steam Tractor

Fig 212: Ruston Steam Caterpillar for export to the Yukon

Frederick Savage

St Nicholas Iron Works, Kings Lynn, Norfolk

Frederick Savage - born in Hevingham, north of Norwich in 1823 and spent his young childhood bird-scaring for a few shillings a week. Apprenticed to Thomas Cooper, implement maker in 1844, he later worked for the future traction engines builders, Holmes & Son at Prospect Works, Norwich. In 1851, he became foreman at Willetts, Baker St, Kings Lynn and when they closed in 1853, Savage set up his own business specialising in the repair of hand-operated winnowing machines. He operated from a workshop in the yard of the Mermaid & Fountain Public House, Tower St, Kings Lynn. Joined by his brother Will, "F Savage, Engineer, Kings Lynn", moved to larger premises built on the site of the old workhouse and christened St Nicholas Iron Works. The first portable engine was built in 1862 and this was later fitted with chain drive to make it self-moving. The business moved to larger premises to the north of the town but kept the same name. The first chain drive engine, no 112, was built in the same year. The last built to this design, no 157, was built in 1876, and the very last chain engine, no 309, was completed in 1883, at the same time as the first of the gear driven machines. From this was derived the famous range of *"Sandringham"* engines available in three sizes, 6, 7, & 8 nhp. In turn these formed the basis for a similar range of road locomotives of heavier construction. Back in 1876, Savages produced their *"Agriculturalist"* engine fitted with cast iron rear wheels. Tread plates on each of these wheels could be removed to expose a 9" wide x 6" deep channel carrying 1000 yards of wire rope. If the rear of the engine were jacked up and supported on blocks, it could be used to power a roundabout system for ploughing. In 1894, the firm experimented with a unique design of compound featuring an annular layout with the low pressure cylinder enclosing the smaller high pressure cylinder. The first engine was built as a showmans road locomotive and the company went on to produce a whole range of equipment for the fairground trade, including the "centre engines" for roundabouts. The most famous of the Savage showman's engines are the 3 combined centre-traction machines for which Frederick Savage obtained a patent in 1880. He became the mayor of Kings Lynn shortly before his death in 1897. The company went on to produce between 25 and 35 steam wagons to a total of 5 designs, all of undertype pattern and using water-tube boilers. The first of these appeared in 1903. The following year, they ventured into the steam tractor market but produced fewer than 10 of the *"Little Samson"* single cylindered 5 tonners.

Fig 213: Savage Thresher c 1870

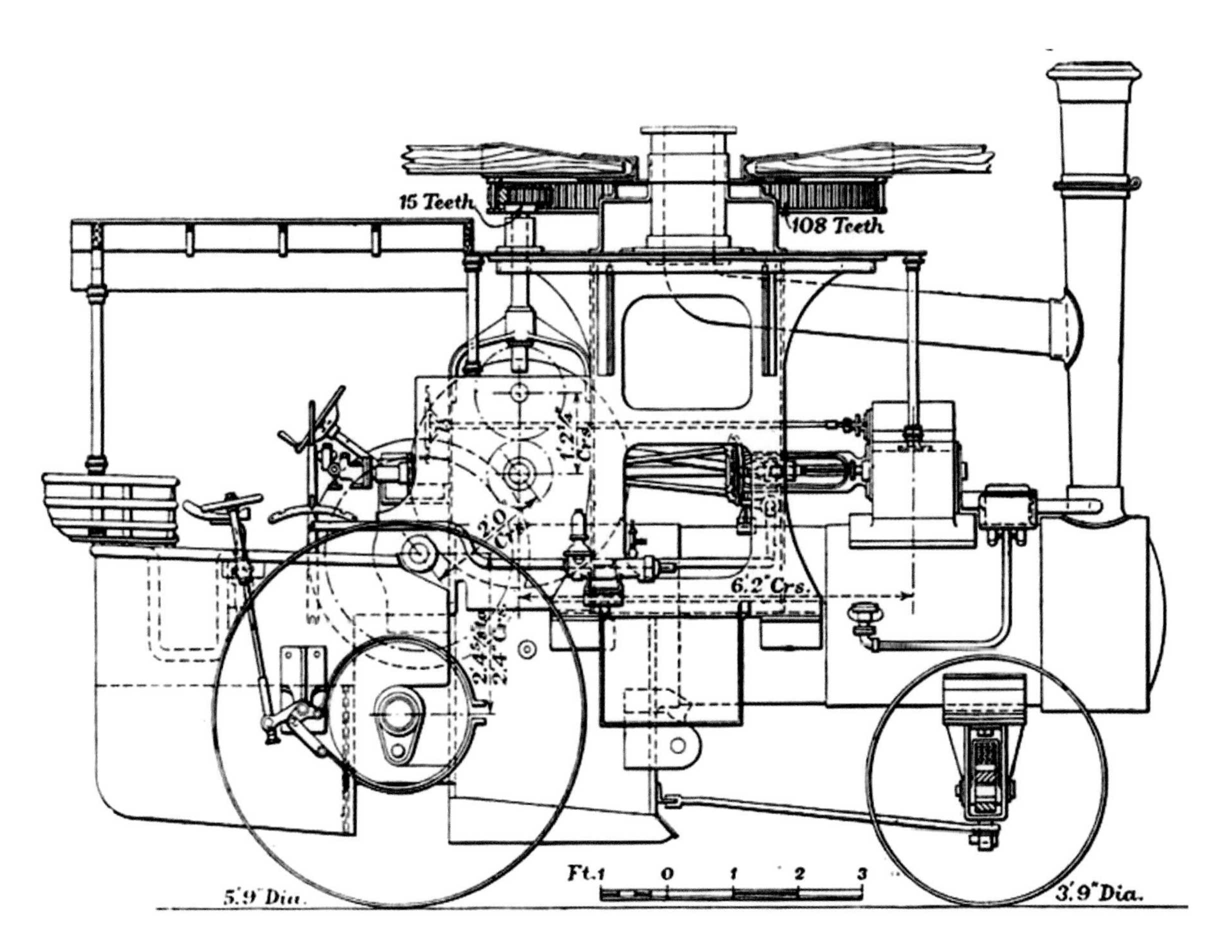

Fig 214: Savage Showman's Centre-Engine

Taskers

Waterloo Iron Works, Abbotts Ann, Anna Valley, Andover, Hampshire

Taskers of Andover was founded by journeyman blacksmith, Robert Tasker in 1814. As a Quaker, he opposed the Napoleonic War and did not receive the support he perhaps deserved from the local community, particularly, the magistrates. In 1830, a 300 strong mob attacked his factory, their main target being threshing machines. A set of roundabout ploughing tackle was built in 1862, with the first portable appearing in 1865. The first traction named "Hero", was built just 4 years later, its front-steered design continuing until 1878, with the first compound appearing in 1894. Taskers were never a major producer of traction engines but their steam tractors were considered by many to have been the best on the market. A few 3 ton machines were built before the change in the law increasing the maximum weight to 5 tons. The 300 or so Roller-Chain drive "Little Giant's alone accounted for over a third of steam engine output. Easily the most famous of these was 1298. Named "The Horses Friend", it was bought by a Mrs Grady and a Miss Perry for assisting horse-drawn vehicles on the Crystal Palace area of London. The engine is currently in preservation. One unusual machine produced by the company was Diplock's "Pedrail" design for an all-terrain machine which was almost a return to the Boydell designs of half a century earlier.

In total the company produced over 1150 threshing machines and were responsible for a number of design patents between 1857 and 1867. These included adjustable inclination for the riddles, parallel motion for the shakers, and an integral chaff blower. Prizes were won at a number of shows in Yorkshire in 1865, Bury St Edmunds in 1867, and the Channel Islands in 1871 to name but 3. The earlier machines were of a single-blast design, the last of these appearing in 1884. They were replaced by a double-blast arrangement, the last of these coming off the production line in August 1923. The company also built over 2000 elevators to a range of patterns starting with horse drive and later incorporating an ingenious folding arrangement.

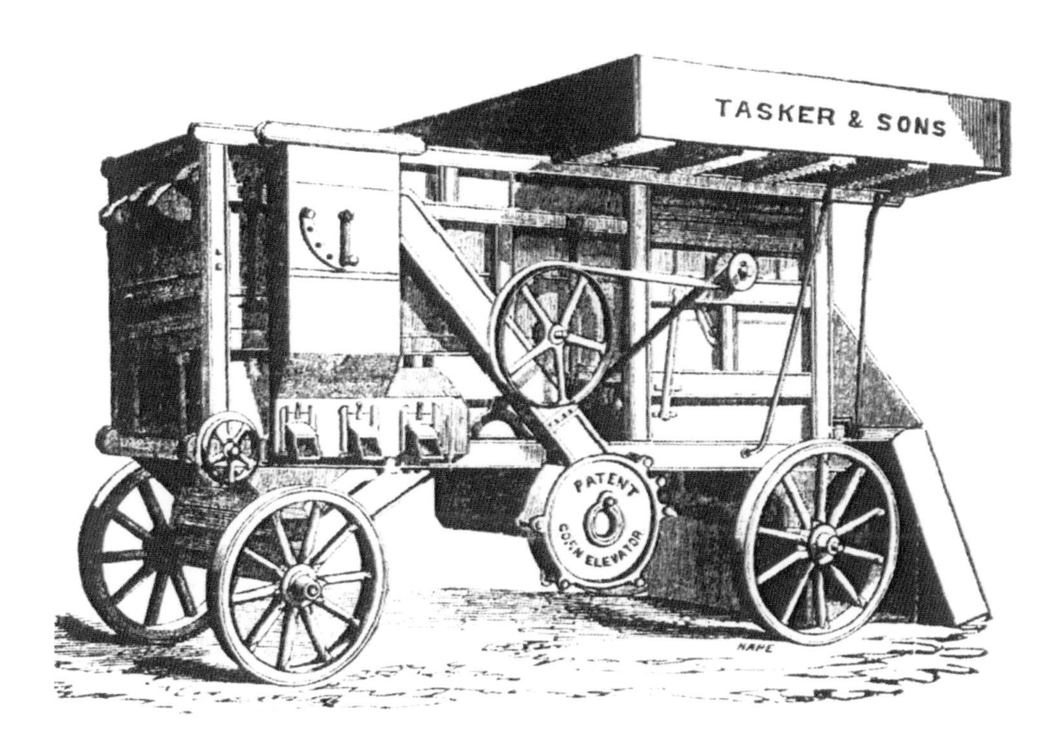

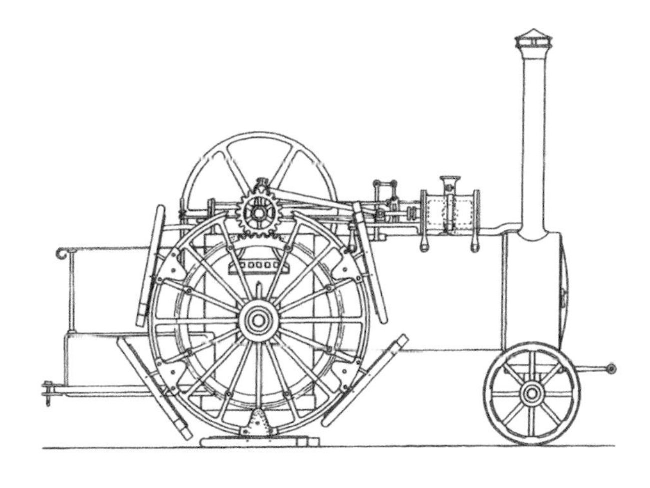

Fig 215: Taskers Thresher c 1870

Fig 216: Taskers first Traction Engine

Fig 217: Taskers "Little Giant" Steam Tractor

Fig 218: Diplock's "Pedrail" of 1902

Fig 219: Boydell Engine by Richard Bach of 1854

Wallis & Steevens

North Hampshire Ironworks, Basingstoke, Hampshire

Richard Wallis & Son had started a foundry 4 years before the North Hants Ironworks opened in 1853, and Charles James Steevens, a fellow Quaker joined the company in 1859. It should be mentioned that the Quakers were often unpopular particularly because of their pacifist views which prevented them from supporting the Napoleonic War.

The firm sold a large number of threshing machines, the earliest being horse-driven before turning over to the use portable engines. The earliest threshers and Portables were built by Clayton & Shuttleworth, before the company built their own machines in 1861 and 1867 respectively.

The firm was joined by designer William Fletcher whose first project in 1872 was a redesign for their range of portables, starting with a 8 nhp that they exhibited at the Smithfield Show of that year. The first traction engine appeared in 1877 and was exhibited at the Bath & West Show of a year later not long before Fletcher left the company to join Burrells and shortly afterwards, Marshalls. A traction convertible to a roller was built in 1890 shortly followed by their first conventional roller, giving some indication as to the direction the company was to take in later years. The most significant design innovation in their traction design came in 1894 with the 2 patents for expansion gear. The first steam tractor, a 3 tonner, was built as early as 1897,the design being subsequently updated in order to take advantage of changes in legislation. This experience with small steam engines formed a basis from which the company were able to develop the lightweight "Advance" and "Simplicity" rollers, and from which they went on to build a range of diesel-engined machines. Wallis & Steevens built a total of 127 steam wagons and unsuccessfully sued Fodens for patent infringement. By comparison, Foden's total wagon output was to exceed 7,000.

One writer has described Wallis & Steevens' thresher designs as " keeping abreast of development but breaking no fresh ground". This did not prevent the machines from winning prizes at the RASE Shows at Canterbury in 1860, and Bury St Edmunds in 1867, plus at the Austro-Hungarian Empire Exhibition held in Vienna in 1873. The first of these used a Brown & May portable to drive the machine, but after that, they built their own.

Fig 220: Wallis & Steevens Thresher c 1910

Fig 221: Wallis & Stevens "Expansion" Engine

Fig 222: Wallis & Steevens Nameplate

Fig 223: Wallis & Stevens "Advance" Roller

Fig 224: Wallis & Stevens "Simplicity" Roller

Barrett, Exall, & Andrews

Kategrove Ironworks, Reading, Berkshire.

The firm built hand-operated machines from early 19th century, and continued to do so into the early 20th century. Accounts of RASE show of 1847 mention a "2 horse thrashing machine with bevil (bevel) gear work" built by the company. A more recent design of 42" sized machine was quoted at £48.00

As with many of their competitors, they built a number of Portables, one of which they exhibited at 1851 Great Exhibition. This featured a type of expansion gear where the governor controlled a link motion, but this prototype did not go into production.

William Brown from Horton Devizes, and Charles Neale May from Basingstoke, join as apprentices in 1846 aged 16, and went on later to found the famous company, Brown & May based at the North Wiltshire Foundry, Devizes.

Fig 225: Barrett, Exall, & Andrews Portable

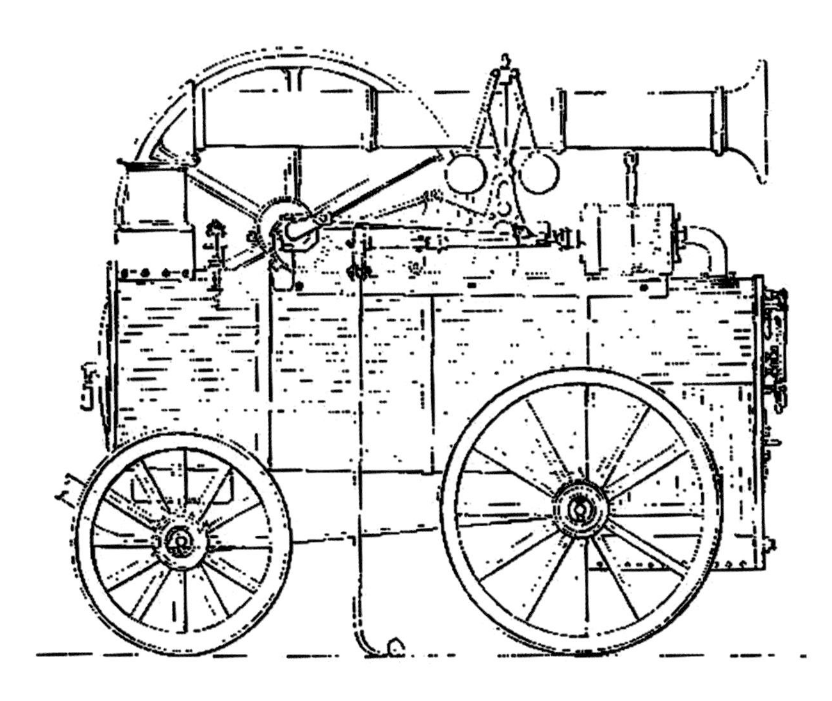

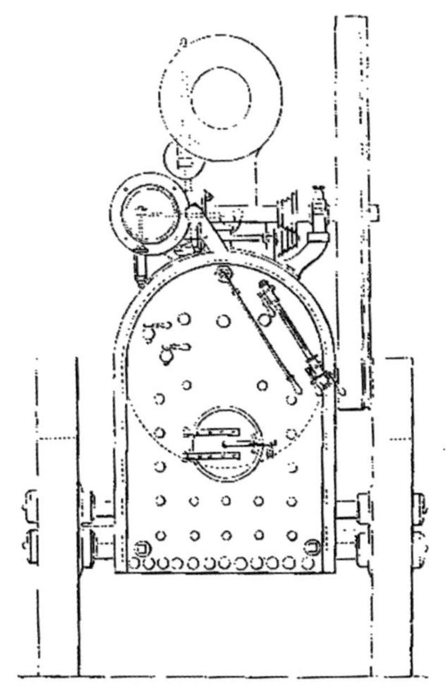

Barrows

Banbury, Oxfordshire

Also known as: Barrows & Stewart, Kirby & Barrows, Barrows & Carmichael.

The company produced a number of threshing machines plus portable engines to drive them exporting a fair number as "threshing sets". The firm moved into traction engine manufacture for a short time, exhibiting examples at the Smithfield Show in 1879 and at the Derby RASE show two years later.

Fig 226: Barrows Thresher c 1910

Fig 227: Barrows export Portable

Fig 228: Barrows Traction Engine

Davey-Paxman

TELEGRAPHIC ADDRESS
"PAXMAN, COLCHESTER."
TELEPHONE No 52.

LONDON OFFICE:
78, QUEEN VICTORIA ST., E.C.

CODES USED: A.B.C. (4TH AND 5TH)
A1, ENGINEERING (1ST AND 2ND)
MOREING & NEAL
LIEBERS & PRIVATE.

DAVEY, PAXMAN & Co. LTD

ENGINEERS AND BOILERMAKERS,

COLCHESTER,
ENGLAND.

19

MAKERS OF
PORTABLE ENGINES.
TRACTION ENGINES.
UNDERTYPE ENGINES.

REFERENCE
YOUR
OUR

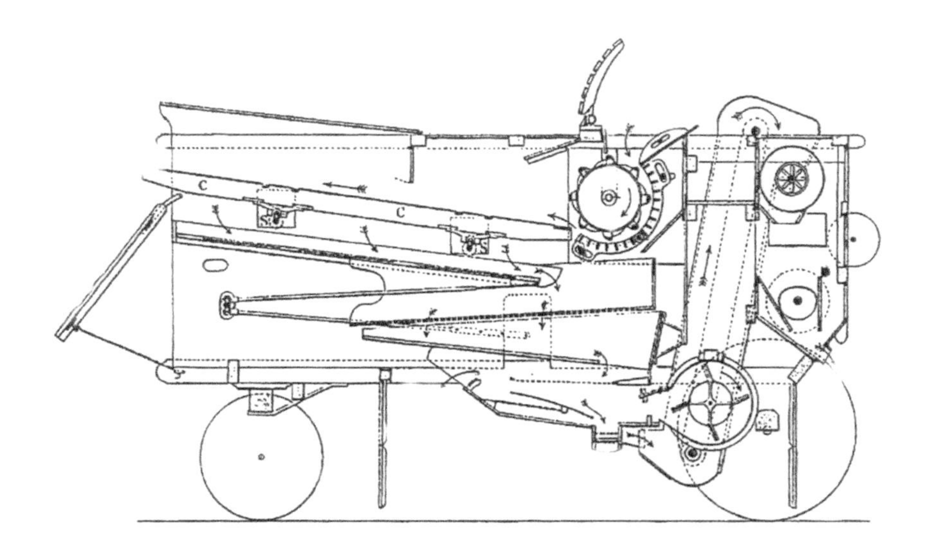

Fig 229: Davey-Paxman Thresher c 1880

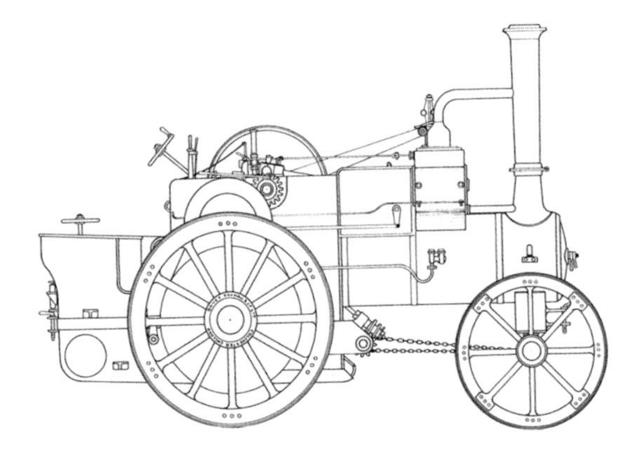

Fig 230: Davey-Paxman Traction Engine

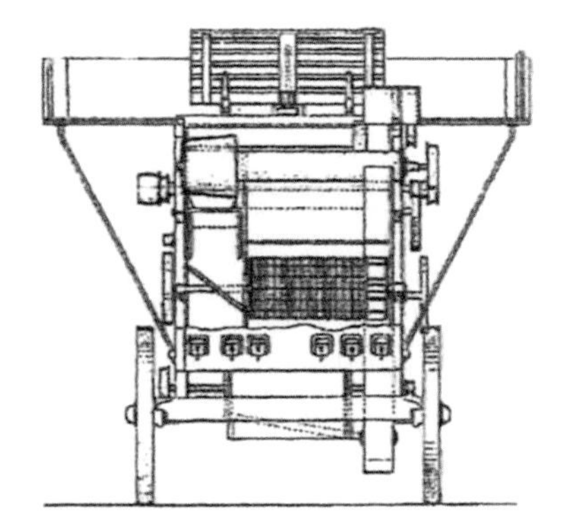

Fig 231: Davey-Paxman Thresher c 1880

Fig 232: Davey-Paxman Digger

P. & H. P. Gibbons

Wantage, Oxfordshire.

RASE records show a Gibbons machine being exhibited at the rain-soaked 1879 Kilburn Show.

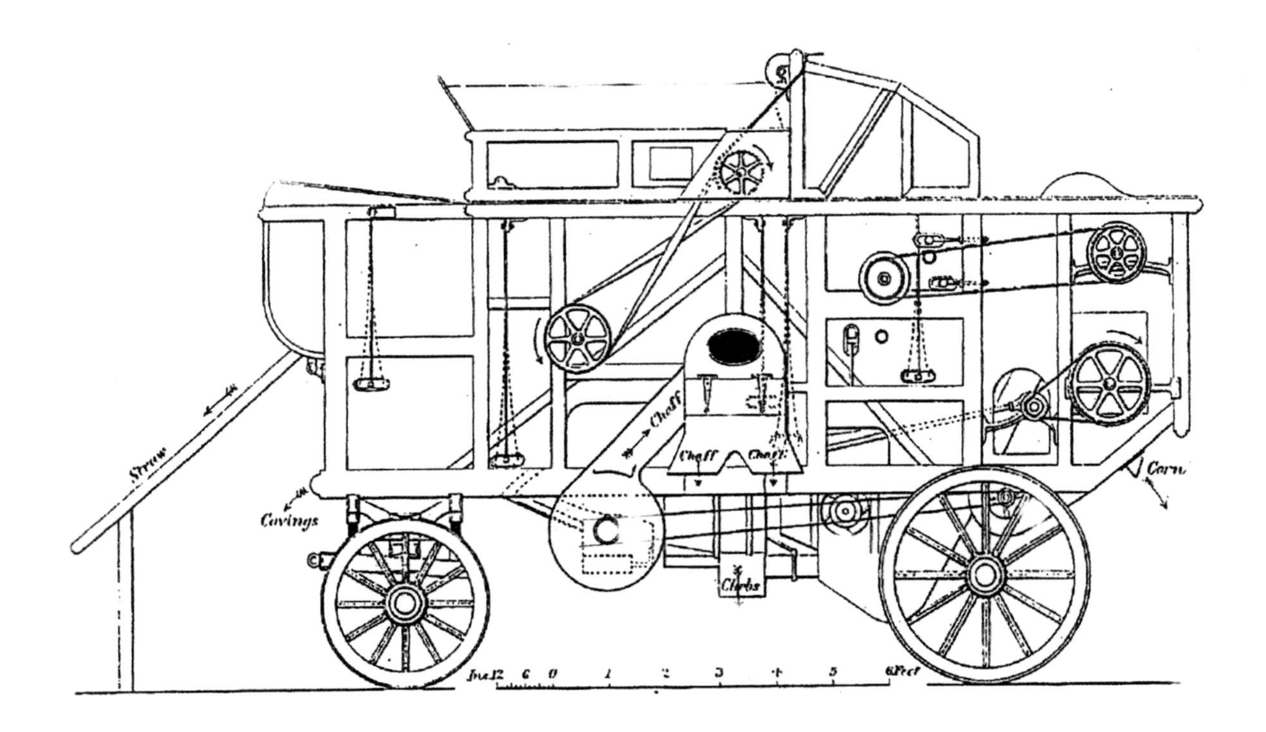

Fig 233: Gibbons Thresher c 1880 -Elevation

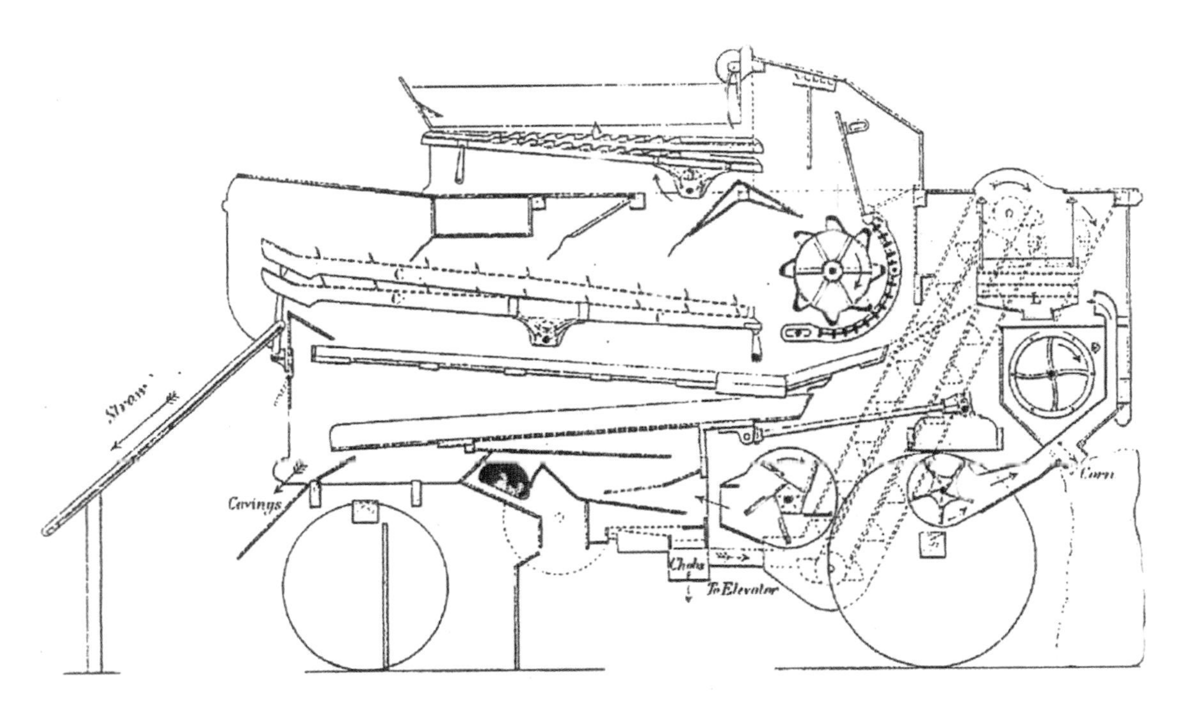

Fig 234: Gibbons Thresher c 1880 -Section

Edward Humphries & Sons Ltd

Later known as: Fisher, Humphries & Son, Bomford & Evershed, Horsfall Destructor Ltd.

Atlas Works, Pershore, Worcestershire.

Humphries were the largest producers of threshing machines in England not to build steam engines commercially, although even they constructed a small number of portables early in the firm's history. They were a much smaller firm than rival manufacturers such as Ransomes or Clayton & Shuttleworth but were able to maintain steady sales over a lengthy period. Their most distinctive design feature was their so-called "humless" octagonal drum. The early designs of many manufacturers featured the dressing apparatus mounted at 90 degrees to the rest of the machine. This arrangement was retained by Humphries throughout their production, and it is interesting to note that when the early combine harvesters arrived, they too were all built to this same layout.

Fig 235: Humphries Thresher 1875

William Tuxford & Sons

Skirbeck Ironworks, Boston Lincolnshire.

The firm was founded in 1826 when the Skirbeck factory was built. In 1830 William Tuxford obtained patent no 5954 for a machine for "reeing" wheat. This operated on the principle that rapidly moving sieves caused lighter material to rise to the surface where it could be removed by hand. They produced one of the very first steam-powered threshing machines in 1842, a design that placed thresher and engine on the same chassis. In common with a number of other companies, they built an engine using Boydell patent wheels. For many years they built traction engines to a very unconventional design using their enclosed vertical "steeple" type motion. Their later design of engine from 1871 onwards used the accepted layout of horizontal cylinders and motion on top of the boiler but the machines ran "tail first". The company finally closed in 1887.

Fig 236: Tuxford Threshing Set

Chapter Ten

ANCILLARY EQUIPMENT

Reapers & Binders

Massey-Harris Binder Poster

Logic dictates that a book about threshing ought to at least mention reapers & binders rather in the same way as Mrs Beeton wrote in her famous cookery book "first catch your rabbit". The reaper, and its later derivative the binder, was the "OTHER" piece of equipment that brought mechanisation to the harvest. The earliest attempts to harvest grain mechanically were made by the Romans. The vallus was a large comb between a pair of wheels pushed by a horse in shafts. The ears of corn scraped off by the combing action were raked off the machine by hand for further treatment.

Fig 237: Roman Vallus

The first modern reaper was designed by a Scottish clergyman, Rev Patrick Bell in 1826. Like the Roman Vallus, it was carried on a pair of wheels pushed by a horse, or to be more accurate, two. A reel rotated above a pair of reciprocating bars carrying a number of pointed knives. Behind this was mounted a transverse canvas which delivered the crop to one side. Drive to all the moving parts was taken off the wheels. It was capable of a work rate of 1 acre per hour, and the Highland and Agricultural Society awarded it a £50.00 prize. Reports state that 10 machines were in use in Scotland by 1832 and some were exported to USA and Australia. But the state of the country at this time meant that labour-saving devices in agriculture were not in great demand. The economic conditions in America were totally different and the history of development of the reaper/binder move to North America where it became that of the rivalry between two men, McCormick and Hussey.

Cyrus Hall McCormick of Virginia, then aged 22, first demonstrated his reaper in 1831 but did not patent it until late in 1833. This featured the familiar row of V shaped blades or knives fitted to a reciprocating bar and a row of projecting "fingers". Earlier in that year the American Quaker, Obed Hussey patented his famous reaper design. McCormick opened his factory in Chicago in 1847 and was soon producing 500 machines a year. Both McCormick and Hussey took their machines to Europe and showed at the Great Exhibition of 1851. Both were horse driven but pulled by horses rather than being pushed as in Bell's earlier design. Neither machine incorporated the canvas delivery system of Bell, all the material having to be raked off the cutter deck by hand. Maximum work rate for both machines was around 20 acres per day. In the trials, the McCormick design won first prize, being better able to cope with the damp English conditions. Other manufacturers were quick to adopt the new principle. In 1852, the British firm of Garretts demonstrated their own reaper based on Hussey's design. In the RASE trials at the Lewes Show, it was reported to have "outrun all its competitors" to win first prize. In the very next year, reapers built by both Garretts and Ransomes were redesigned to incorporate an elementary form of sheaf gathering system. The year of 1853 saw no fewer than 1500 such machines produced by British manufacturers. In the 1853 trials organised by RASE in Gloucester, Henry Crosskill of Beverley using an improved version of Bell's original design triumphed over opposition which included all the foremost American designs. 1854 saw the Dray company take first prize at the RASE Lincoln trials again using a Hussey-based design, but Burgess and Key won the following year with a McCormick-based machine. Despite Bell's use of a canvas in his first design, other manufactures were slow to replace the hand-raking off the cutter deck by mechanical means. An 1854 design by Burgess and Key of Brentwood featured an archimedean screw but in 1856, the American Owen Dorsey exhibited a machine fitted with a series of rake arms rotating on a vertical axis.

Fig 238: Sail Reaper

From 1860 onwards, thousands of machines were built to this design which became known as a sail reaper. A number were built in Britain and one by Samuelson & Co of Banbury won first prize in the RASE trials of 1862. With their 6' cut, these machines were capable of a work rate of over 30 acres per day. Back in America, McCormick had designed his first self-rake reaper known as "old reliable" in 1858. A single reaper required 5 people to follow the machine gathering the crop and tying it into sheaves. Charles Marsh of Illinois found that the same job could be done by 2 people riding on a machine fitted with a canvas delivery belt and patented his "harvester" in 1858. Demand for such machines greatly increased during the labour shortage caused by the American Civil War which lasted from 1861 to 1865.

The next step towards a fully automatic binder was taken by John E Heath of Ohio who devised a mechanism in 1850 to feed twine to the bundles and to cut it to length. This was more efficient than the harvester but still required a second operator to manually carry out the knotting. In 1856, C A McPhitridge of Missouri patented an automatic wire-tying mechanism using a simple twist action, and this system was improved upon by Alan Sherwood in New York in 1858. This use of wire ties carried with it the risk of injury to livestock from ingesting the short lengths of wire. It was not popular with many farmers and in the USA, the problem being euphemistically referred to as "hardware disease". Concerns were also expressed about possible damage to threshers and mills, not to mention pieces of wire appearing in loaves of bread. Some modern balers still use wire. For purposes of comparison, modern baler wire is annealed 14½ gauge with a tensile strength of between 50,00 and 70,000 pounds per square inch and most tying mechanisms give 3 twists.

A number of designs were produced to tie the bundles with twine or even a length of straw. John Appleby of Wisconsin developed one of the first effective knotter mechanisms but it was Jacob Behel of Rockford, Illinois, that produced a knotting arrangement based on the "billhook" principle giving the "round knot" that is still in use today. In 1874, a Mr Witherington further developed the binding mechanism Although this development work was taking place in the USA, specific sheaf-binder trials were organised by the RASE in the UK. A machine built by Walter A Wood received a notice at the 1876 Birmingham Show. For the first time the trials were held away from the showground, this time near Leamington Spa. The following year at Liverpool Show, no prize was awarded. By the Bristol Show, a year later, 4 wire-tying machines were competing against 3 string-tying Binders. Prizes were awarded to Burrells, Waite and Huggins & Co using McCormick's design. The RASE judges were anxious to promote these safer twine machines from the early 1880's onwards. By this time, a number of American manufacturers had begun using twine made from a mixture of Mexican sisal and Manila hemp. Flax-based twine had been tried briefly but was found to be too prone to vermin attack. Binders to this design were produced by a number of British manufacturers including Bamlett, Samuelson, Hornsby and Howard. This was in addition to the many McCormick machines imported directly from the USA. Their first self-binder was produced in 1876 and this evolved into the twin twine machine he patented in 1881. This was so successful that McCormick discontinued production of wire tying machines completely in 1883. By 1900 over 75% of British corn cut was by binder and McCormick's Chicago factory was already turning out 4000 machines a year. Peak of binder use occurred in 1950 with 150,000 machines in Britain alone. From this date on numbers fell as a result of the changeover to combines harvesters. As an aside, the simpler hay mower developed in the 1850's used the technology developed for the reaper.

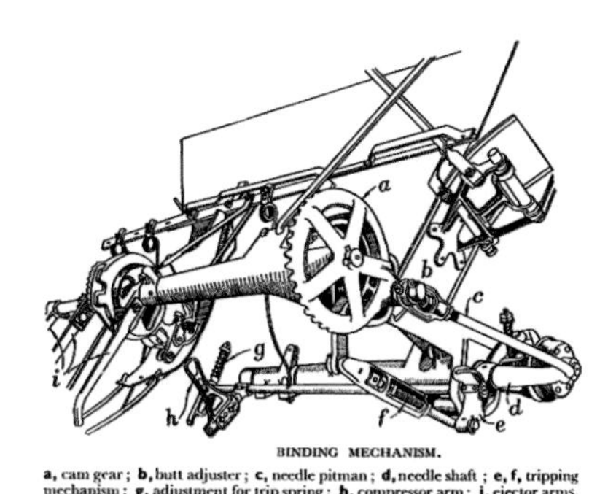

Fig 239: Binder Mechanism

Fig 240: Modern pattern PTO-driven Binder

Elevators

Before elevators, pitching straw to higher ricks involved a chain of men working off a number of intermediate platforms. The elevators that replaced this method of working may be divided into two different groups: undertype and overtype.

The undertype, usually called farmers' or "stacker" pattern, is a multi-purpose machine consisting of a wide endless belt of wooden cross laths on a pair of chains running over a flat platform with low sides. It is suitable for most materials but not ideal for loose straw, which is easily blown away by the wind. The elevator may be a single unit or of a two-piece design. Some are telescopic, whilst others are hinged. Hinged models either fold back on themselves and travel in lowered position, or fold forward and travel in raised position.

The first overtype was built by Edward Hayes, Watling Works, Stony Stratford in 1867. Here the straw is taken up a fixed trough by a series of overhead metal rakes which return over the top. The main advantage of these machines is that the layout prevents losses in windy conditions. On the down side, they are unsuitable for other uses and they do not extend.

Fig 241: Early Elevator (Taskers)

Fig 242: Late Elevator (Beare)

The Hayes machine exhibited at the 1876 Birmingham Royal show was priced at £54. These elevators are often referred to as Hayes pattern even when built by other manufacturers. They were available in variety of lengths between 18' and 30'. 53 of these popular machines were built between 1885 and 1889, and a total of 369 had been sold before production ceased in 1951.

Both types of machine generally used a timber frame on 4 wheels including a steerable fore-carriage. Weighing 1½ to 2 tons, they were driven off a 4" diameter pulley on one end of the rear shaker crankshaft. It was generally agreed that they operated best at slower speeds, the slower the better, and some threshing machine manufacturers even offered a smaller 3" diameter shaker crankshaft pulley for use with lighter crops. Drive was via a buckle-jointed crossed 2" belt running over jockey pulleys on angle irons. This enable the elevator to work at up to 90 degrees to the drive from the thresher. Machines were later driven by a small stationary-type internal combustion engine of about 1½ bhp. Some of these were factory fitted, whereas others were conversions carried out by their owners. A hopper was fitted at the base for feeding. The tension of the twin chains was adjusted by sliding the bearings for top shaft in their mounts, rather like the system used on thresher elevators. The delivery height could be adjusted, either by means of long curved racks, or by a system of pulleys with wire rope on twin side masts. They would operate up to an angle of 60 degrees.

Straw Presses / Balers

Fig 243: Early Baling Press (Harris-Rylatt)

The earliest form of baler was referred to as a baling press. A machine built by Fosters to a design by Edward Sanders in 1888, patent no 5282, was capable of processing 1¼ to 2 tons of straw an hour. This was pressed to density of 25 lbs/cu ft and used hand-inserted "dividing boards" to feed the twine used for tying. Adjustments on the machines allowed hay bales to be produced weighing up to 2 cwt each.

Fig 244: Late Stationary Baler (Jones)

The balers used in conjunction with threshing machines are described as stationary type. These are very different to the machines used today for taking material directly off the ground and are known as "pick-up" balers. A stationary baler takes the straw from above. A feeding arm working in a short arc forces the straw down into a square section chamber where it is compacted by a horizontal piston. Both these components are driven from throws spaced 180 degrees apart on a large crankshaft carrying a heavy flywheel. To avoid damage to the machine in the event of a jam, it is customary to provide a shear bolt in the drive train between drive pulley and flywheel. The density of the bales produced is adjusted by a pair of large wing nuts tightening against a plate that forms the top of the compacting chamber. As with the trusser, the flow of straw is not constant so to keep the bales the same length, they need to be monitored. This is performed by a spider wheel turned as the bale material passes beneath it. A small plain wheel rotated by the spider wheel rolls along the edge of a curved arm until it reaches an indentation. This allows the arm to move triggering the double knotter mechanism and then resetting itself. The bale length is adjusted by sliding a stop block on the end of the arm and locking it in position using a large knurled knob. The Arbor Machine Co of Ann Arbor, Michigan, claimed their 1904 baler could handle 68 tons of hay in a 19 hour working day. As an aside, the company T E Pilter of Paris produced a machine in 1870 capable of producing round bales. These were destined for use by the French cavalry.

Chaff Cutters

Fig 245: Early Barn Chaff Cutter

Fig 246: Intermediate pattern Chaff Cutter

The earliest English patents for a chaff cutter or "chaff engine" were granted to Mr J Cooke of London, and Mr W Naylor of Langstock, in 1794 and 1795 respectively. A modified design by Robert Salmon of Woburn in 1797 received a prize of 30 gns from the Society of Arts. This consisted of a pair of large diameter wheels with 2 spring-mounted blades set at 45 degrees between them. The wheels operated a crank that gave the pair of feed rollers an intermittent motion using a pawl-and-ratchet system. This meant that the straw was held still as the blades performed the cutting action. A series of 4 holes on a block mounted on one of the wheels plus another 5 holes at the end of the crank, meant that the throw of the crank could be varied providing no fewer than 20 different lengths for the cut material.

An improved design by Thomas Passmore of Doncaster became known as the "Doncaster Engine". Here the principle remained the same but the diameter of the wheels was greatly reduced.

Mr W Lester of Paddington, already mentioned for his improvements to thresher design, produced his design of chaff cutter in 1801. Known as the "Lester Engine", it was on this basic design that machines were still being produced in the 1840's. The machine produced by Thomas Heppenstall of Doncaster was rated at an output of 15 to 20 bushels an hour. Yet another design was developed in the 1840's jointly by a Mr Budding and Mr R Clyburn who had produced the improved fan blade profile used for winnowing on threshing machines. At this time, both men worked for Lord Ducie on his experimental farm at Whitfield, Gloucestershire.

The machine in production by Ransome and May in the 1840's to Mr C May's patent delivered 300 to 350 cuts per minute to straw fed through an opening of 30 to 50 square inches allowing 12 cwt of straw or hay to be cut into ½" lengths in an hour. The machine was powered by 2 horses.

The original hand-operated chaff cutter used a pair of toothed rollers to control the feed thus avoiding chokage. The upper roller was arranged so as to be able to rise to accommodate thicker bundles. The feed opening was 9" wide by 1½" to 4" high. A heavy flywheel carried two curved knives angled slightly inwards giving a continuous and uniform cut with the rotation of the flywheel. The length of the cut pieces could only be altered by changing a pair of gear wheels. For use as fodder, the straw was usually cut into ½" lengths. All these later machines used a simple system of gearing to the pair of feed rollers. This gave a continuous feed, as opposed to the intermittent motion of the earlier designs.

Chapter Eleven

HISTORICAL CONTEXT

RASE Seal

It is worth some digression to put the main contents of this work in a historical context. The growing, and more importantly as far as this book is concerned, the attendant threshing of cereals is one of man's earliest activities. The earliest of records of cultivation relate to the area known as the fertile crescent which includes present day Iraq and Iran. Much of this cultivation was of cereals so there would have had to have been threshing. Ancient Egyptian cereal production is reasonably well-documented and includes mention of the traditional threshing sledge, the charatz. The Pharoah is often depicted holding a stylised flail as part of his regal insignia showing the importance of this activity. A little later the Bible makes reference to the Hebrew equivalent, the morag and later still, Latin texts refer to the Roman tribulum. Despite these early attempts at simple mechanisation, by far the majority of threshing was carried out using the flail, and this persisted in Britain until comparatively recent times.

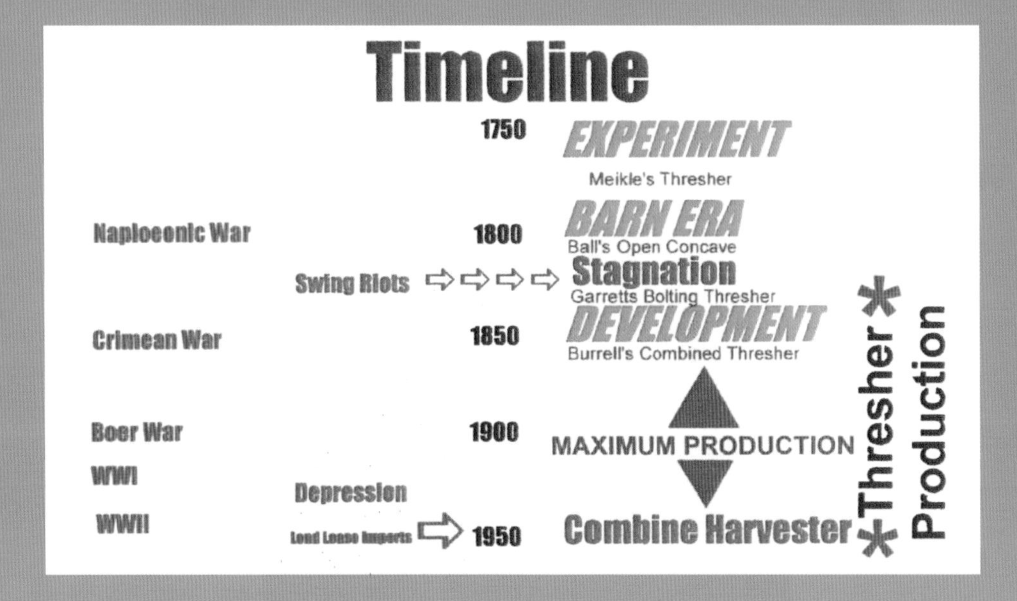

Fig 252: Timeline

In the mid 18th century, following research by Jethro Tull, many more cattle began being kept over the winter rather than being slaughtered. This required the growing of very much more material both for fodder and for bedding. Tull's design for a seed drill was one of the first major steps in the mechanisation of agriculture and this coincided with the invention of the flying shuttle by John Kay in 1733 which heralded the Industrial Revolution. It was the textile industry that was first to introduce large-scale mechanisation and over just a few years, it changed beyond all recognition following a whole list of inventions:The Spinning Jenny by James Hargreaves in 1764, the Water Frame by Richard Arkwright in 1769,the Spinning Mule by Samuel Crompton in 1779, and the Power Loom by Edmund Cartwright in 1785. Much of this and future development was made possible by the adoption of the steam engine as the principal power source. James Watt had produced his first steam engine in 1760 based on the pioneering done work by Thomas Newcomen some 50 years before.

In 1732 Michael Menzies, obtained the first patent for a power-driven threshing machine and the first machine that could be described as anything like successful was produced by Leckie about 1758. The much better-known Andrew Meikle built his first of two less successful prototypes in 1778, and the well-documented thrasher was patented 1788, manufacture commencing in 1789.

What also began in 1789 was the French Revolution which alerted the English aristocracy to the distinct possibility of a repeat performance in Britain. The nearest this got was an Irish attempt at independence. Interestingly this was with the active, if somewhat ineffectual support of the French. After the Revolution came the rise of Napoleon whose aspirations of a European empire drew Britain into the Napoleonic Wars.

Fig 253: "Promis'd horror of French Invasion"

The period immediately following Napoleon's defeat at Waterloo was a troubled time for English agriculture. The first general Income Tax had been introduced in 1799 to pay for the war, and this resulted in much lower wages. Early nineteenth century England was virtually unique among major nations in having no class of landed smallholding peasantry. A series of Acts of Parliament collectively known as the inclosure Acts removed the rights of country folk to graze their livestock on what had formerly been common land. This grazing area was divided up among the large local land-owners, leaving farm workers landless and entirely dependant upon offering their labour for a cash wage. The earliest of these Acts date from the Middle Ages, but these were few in number. A whole series began around 1750 and in 1801, the First General enclosure Act was introduced. This was designed to simplify the procedure which effectively overrode any local objections. Further Acts followed in 1821 and 1836 until finally in 1845, a Second General Enclosure Act appeared which speeded up the process still further. This was achieved by using appointed commissioners to designate the areas to be enclosed, which meant that it was no longer necessary to refer individual schemes to Parliament. Rising corn prices led a number of landowners to meet at the Pelican Inn at Speenhamland near Newbury on 6 May 1795. The outcome was the so-called Speenhamland Act which allowed the wages of agricultural labourers to be subsidised from the poor rates, the payments being dependant on family size, a sort of Georgian "family credit arrangement. This offered a solution during the boom years of the Napoleonic wars when labour had been in short supply and corn prices high, but the coming of peace in 1815 brought with it an oversupply of labour. The number of war veterans returning exceeded, 200,000. Plummeting grain prices produced a corresponding reduction in the funds available for the poor rate.

Central to British eating habits has always been the consumption of bread. The Magna Carta of 1215 sought to bring some regulation and proclaimed "there shall be but one measure throughout the realm". This was formalised by "The Assize of Bread", Assisa Panis et Cervicae established in Bury St Edmunds of 1266, and this was to remain unchanged until well into the 19th century.

If three and a half "one gallon" loaves had been considered necessary for a man in 1795; by 1817 the provision had fallen to just two. The size of loaves was controlled by the first of two Bread Acts in 1822. The Poor Law Act of 1834 established the workhouses, a system that lasted all but a century, not ending until 31 March 1930. This Poor Law Act was closely followed by a second Bread Act in 1836, a statute which remained without amendment until 1907.

The falling corn prices led to the Corn Laws, a body of legislation which came into force in direct response to the end of the Napoleonic wars. These were essentially import tariffs designed to support and protect domestic British corn prices against competition from less expensive foreign grain imports. The first tariffs were introduced by the 1815 Importation Act and finally were repealed by the 1846 Importation Act. Britain in the first quarter of the nineteenth century was the most economically developed country in the world. The "protection" provided by the Acts was not so much against imports from foreign countries, as it was against cheap imports from the British Empire. These would have severely cut into the profit margins of British farmers and landowners, and the resulting artificial rise in food prices made the lot of the agricultural labourer even worse. An unarmed demonstration in Manchester advocating the repeal of the Corn Laws in August 1819 resulted in the infamous Peterloo Massacre, where soldiers attacked the crowd killing 11 and injuring an estimated 500.

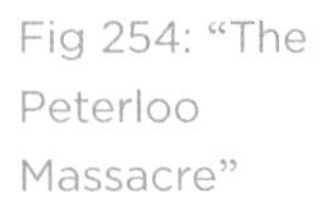
Fig 254: "The Peterloo Massacre"

This government action was calculated to forestall a growing tide of radicalism and further repression resulted from another body of legislation collectively known as the Six Acts. These Acts included such measures as seizure of arms, the outlawing what would now be described as para-military activities, and the banning any organisation that could be in any way "Seditious". It was under this legislation that the well-known Tolpuddle Martyrs were convicted in 1834. These 6 men of Dorset were sentenced to 7 years transportation for forming an illegal union-style organisation named innocuously "The Friendly Society of Agricultural Labourers". Their principal demand had been for a minimum wage of 10/- a week when the offer at the time was 7/-,an offer that was soon after to fall even lower to 6/-. Four years later, more people were prosecuted under these repressive pieces of legislation for demanding what now might be described as a bill of human rights known at the time as the "People's Charter", and giving the group its title, the Chartists. Going back to the Corn Laws, a measure of their effectiveness can be seen when it is realised that Britain's dependence on imported grain in the 1830's was around 2%. By 1860, this had risen to 24%, and by the 1880's, 45%. A report commissioned in 1934 showed that no less than 67% of ALL food in Britain was imported.

One piece of industrial action used by a more militant faction of the Chartist group is particularly interesting in the context of this book. Following an explosion in one of his high-pressure boilers in 1803, Richard Trevithick developed the first fusible plug.

Fig 255: Fusible Plug with lead centre

Fig 256: Not all Fusible Plugs are the same size

If the water level in a boiler is allowed to drop too far, the copper roof or "crown" of the firebox becomes exposed. Above 550 degrees centigrade, the copper can lose strength and fail structurally. To solve this problem, a bolt with a tapered hole through its centre filled with lead or tin is screwed into the "crown" of the firebox. The lower melting-point of the lead or tin means that the pressure is relieved and steam is released onto the fire, probably extinguishing it. "Dropping the plug" was always seen as the mark of the incompetent engineman. The simple expedient of removing the fusible plug from the roof of a steam engine firebox makes the machine instantly unusable, without causing any actual damage. This is similar to removing the distributor cap from a modern car engine. And like distributor caps, these plugs were a purpose-designed component that could not be replaced by anything other than a genuine spare part. This technique gave its name to a series of actions known as the "plug plots".

Two terrible harvests in 1828 and 1829 had prompted the radical journalist William Cobbett to say "it would better be a dog than a farmer next winter" following his tour of Britain on horseback. His concerns were well-founded and farm labourers faced the approaching winter of 1830 with dread. The villagers of Southern England felt their very way of life was facing extinction. Up to then, the threatening of owners of machinery followed by rick-burning and cattle-maiming had been carried out by farm labourers only, but this "last labourers' revolt" was now joined by village craftsmen, all of whom felt their livelihoods threatened.

Threshing machines were very much a priority target in this revolt. The total time spent by farm labourers with flails was enormous. The wheat from one acre of land required around 70 man-hours to thresh by flail. The maximum amount of wheat that a man could flail in a single day was reckoned to be less than 10 bushels. Apart from the sheer amount of potential work lost, the situation was made worse by virtue of the fact that the majority of threshing was carried out during the winter. What little other work was available would have been out of doors in all winds and weathers. So the prospect of steady employment indoors was welcomed and Hardy's Tess of the d'Urbervilles waxes lyrical about the camaraderie of a group of flailing labourers. Hardy's attitude is in sharp contrast to that of Scottish poet Robbie Burns who once described the use of the flail as "the most degrading of occupations".

The situation resulting from the popular unrest became so highly-charged that in some areas, the local magistrates went as far as to direct farmers not to use their threshing machines until such time as the levels of unrest lessened. In just 2 years over 400 threshers were destroyed. Not only was machinery destroyed, some of the newly-opened workhouses were wrecked and battles with the local yeomanry broke out. This was the high-water mark of the various Luddite movements. These are more generally associated with factory life, particularly the textile industry. The word Luddite dates from the 1760's and takes its name from a fictional Leicestershire journeyman called Ned Ludd. The threatening letters to farmers and landowners, were often signed "Captain Swing" hence the disturbances of this period being known as the Swing Riots. These were very much based in the southern counties. The economist James Caird (1816-1892) in his book "English Agriculture", drew a line from the Wash to the Severn to demonstrate a north-south divide, where average agricultural wages were over 25% higher above the line because there was competition with industrial employment.

Fig 254: Map of extent of the Swing Riot Area

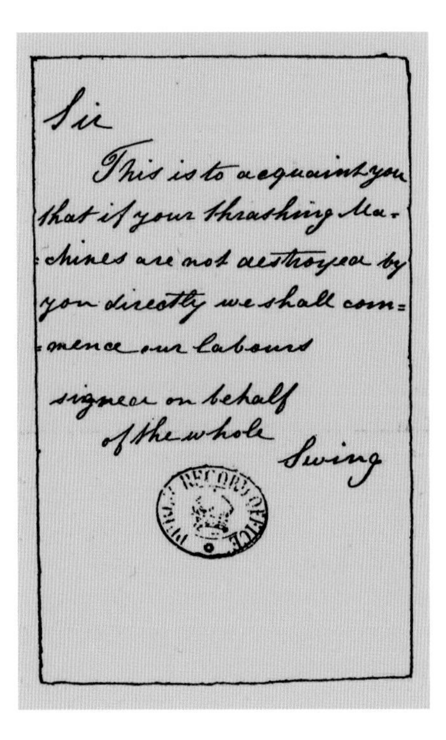

Sir
This is to acquaint you that if your thrashing Machines are not destroyed by you directly we shall com=
mence our labours
signed on behalf
of the whole Swing

Fig 258: Letter from "Captain Swing"

If there was any sympathy shown to the rioters by local farmers, this was not shared by the government, led by the then Home Secretary, Lord Melbourne. As with the troubles in Ireland, the movement was violently suppressed. Nine convicted of riot were hanged and a further 450 had their death sentences commuted to transportation for life to Australia. Of special interest in the context of this book, 2 of those executed and 10 of those sentenced to transportation for life had been part of a 300 strong mob from Andover that attacked Tasker's foundry outside Andover.

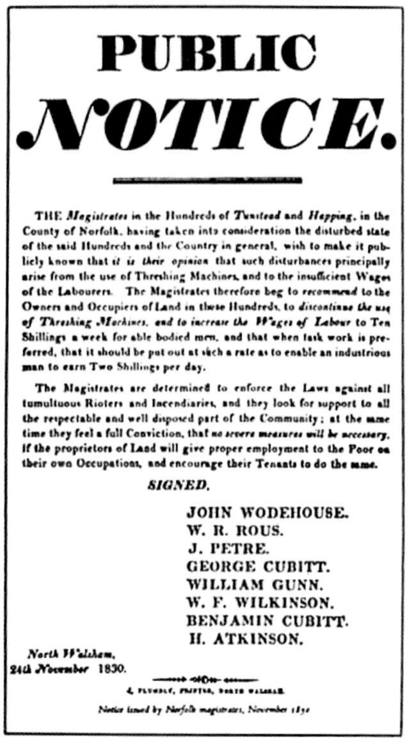

PUBLIC

NOTICE.

THE *Magistrates* in the Hundreds of *Tunstead* and *Happing*, in the County of Norfolk, having taken into consideration the disturbed state of the said Hundreds and the Country in general, wish to make it publicly known that *it is their opinion* that such disturbances principally arise from the use of Threshing Machines, and to the insufficient Wages of the Labourers. The Magistrates therefore beg to *recommend* to the Owners and Occupiers of Land in these Hundreds, to *discontinue the use of Threshing Machines, and to increase the Wages of Labour* to Ten Shillings a week for able bodied men, and that when task work is preferred, that it should be put out at such a rate as to enable an industrious man to earn Two Shillings per day.

The Magistrates are determined to enforce the Laws against all tumultuous Rioters and Incendiaries, and they look for support to all the respectable and well disposed part of the Community; at the same time they feel a full Conviction, that *no severe measures will be necessary*. If the proprietors of Land will give proper employment to the Poor on their own Occupations, and encourage their Tenants to do the same.

SIGNED.

JOHN WODEHOUSE.
W. R. ROUS.
J. PETRE.
GEORGE CUBITT.
WILLIAM GUNN.
W. F. WILKINSON.
BENJAMIN CUBITT.
H. ATKINSON.

North Walsham,
24th November 1830.

[illegible]

Notice issued by Norfolk magistrates, November 1830

Fig 259: Public Notice to a ll Threshing Machine Owners

Away from agriculture, industrialisation and mechanisation were flourishing. The production of pig-iron, the raw material from which all other iron products were derived, saw an elevenfold increase from ¼ million tons in 1801 to 2.7 million tons in 1851. Over the same period, the population rose from 9 million to 15 million, a 67% increase. More significantly, the total urban population rose over this same period from 2½ million to 10 million, a staggering 400% increase.

However this mass migration to the cities had a downside in terms of public health. Apart from the polluted and dangerous conditions in the factories of the time, increasing the population density way beyond what existed in the countryside made disease an even greater menace. After the cholera epidemics of the 1850's, major improvements were made in sanitation but a child born as late as 1900 would only have had a 50% chance of reaching the age of 50. This is a mortality rate equivalent to only the worst areas of the world today, like for example, sub-Saharan Africa. The attendant infant and child mortality rates meant that population numbers were kept stable by a high birth rate. A situation that increased the death rate in women at a time when one in 4 mothers died in childbirth.

Perhaps the area of development most influential in the development of threshing was that of the railways. The first line to take paying passengers, from Stockton to Darlington, was opened in 1825. This made steam locomotion accepted as a norm and, at the same time, fuelled technical innovation. An attempt was made to apply the same principles to road transport in the 1820's in the short era of the steam coach, but this failed dismally. The establishment began a campaign to discourage all mechanised road transport which was to last for the best part of the next 100 years. It was as if the horse or the railway engine were the only acceptable modes of travel. It has often been said that this was because of the vested interests of the railway companies, and whilst that may not have been the only reason, it must have played a significant part. Between 1860 and 1895 no fewer than 15 separate Acts of Parliament were enacted solely to limit the use of powered vehicles on, or even NEAR a road.

As an aside, not only were steam-powered vehicles the target of criticism by the horse-owning upper classes. In 1878, the 'Times' published a number article complaining of "tyranny of the roads". But this time their wrath was directed at the errant behaviour of cyclists.

Quite apart from the potato famine in Ireland in the 1840's, the rest of the United Kingdom entered an era that became known as the "hungry forties". Fortunately, this was followed by the period variously referred to as "Victorian High Farming" or the "Second Agricultural Revolution". The Corn Laws had finally been repealed in 1846 and the social climate in England had stabilised such that the use of the threshing machine and the portable, later the traction engine, rapidly gained favour. It almost seems as if inventive minds occupied in other areas suddenly turned their attentions to the mechanisation of agriculture.

These machines were such an advance that soon many farms had acquired them. Also there was emerging an entirely new breed of users, the threshing contractor. Unfortunately, farm labourers of the time had no real awareness of the potential hazards of the machines and often failed to exercise the necessary care. It should be remembered that in summer, the labourers would be served beer as the only safe drink available. Each man could consume around six pints a day in the hot weather and some farms and mills even had their own brewery. Thresher beater bars travel at over 6000 ft/min or nearly 70 mph and country folk at this time would be totally unacquainted with anything moving at this sort of speed. Not surprisingly, this resulted in a number of serious accidents, not a few of them fatal.

Fig 260: Watercolour of the RASE Show

With an eye for economies to be made, a Mr Lethaby addressing a Burton-on-Trent Farmers' Club meeting in 1843, stated that as steam threshing was less arduous for the labourers than using the flail, a saving could be made in that they required less beer to be bought for them. It wasn't until 1878 that the threshing machine Act was introduced requiring the fitting of some sort of guard or safety feed for the drum but most of the other moving parts were still left exposed. By way of comparison, pre-1914 legislation in Germany required a level of safety guarding not required in Britain until the 1960's.

By 1870, flail threshing had all but disappeared except on smallholdings and for specific purposes such as shelling beans or hulling clover. In 1878, one J Algernon Clark stated that "three quarters of white corn is now reaped by machine". It was Lord Emle that coined the phrase "Golden Age in Farming" and agrarian economists refer to the period between 1830 and 1870 as the "high farming era". This period of agricultural buoyancy gradually came to an end because of another succession of bad harvests from 1875 to 1877, and again in 1879, described as the worst on record. There had also been a series of cattle plagues that had begun in 1867. The disease was known by its German name, Rinderpest,a viral condition akin to measles that usually proved fatal. There followed a period in agriculture from 1879 to 1884 which came to be known as The Great Depression. A rise in cheap American grain imports encouraged many farmers to change from corn growing to pasture and records show that over one million acres were changed over in this way between 1875 and 1885, this figure representing a reduction of over one quarter.

For the manufacturers of farm machinery, home sales of threshers and other harvest equipment suffered badly, and more and more, they were obliged to find overseas markets. Unfortunately their initial success declined. The two reasons usually cited for this by agrarian economists are "slow to adopt technical innovation" and "slow response to foreign competitors". A good example of this would be the domination of North American designs, best exemplified by the binder. The other face of manufacturing was reflected in increased labour costs. Various changes in employment law were appearing such as the 1871 Trade Unions Act which introduced a maximum 9 hour working day. As an example, the management at Fosters were then having to pay a skilled man £2/8/4d for a 53 hour week, equivalent to 11d per hour. Like many of their rivals, they joined the Agricultural Engineers Association in 1875. The whole face of Britain was changing. The 1881 census showed a decline of 92,250 in agricultural labourers since 1871. This was combined with a 53,496 increase of urban labourers, being former farm workers who migrated to the cities to find employment.

The years 1881 to 1884 saw a depression hit the Eastern European export market, and manufacturers were forced to look even further afield. As an example, the firm of Fosters managed to find a market in Argentina where their threshing set, comprising machine and portable, received an "Honourable Mention at the Buenos Aires Exhibition of 1880". Unfortunately political unrest in Central and South America in the 1890's led to a major reduction in sales in these countries too, and the company was forced to look elsewhere once again.

Notwithstanding the above, the development of threshing machines steadily continued and by 1900, the design of the "modern" machine had been finalised. A number of other innovations also emerged including for one, the introduction of the stack elevator in the 1890's. Whilst the home market might have been depressed, the export trade was buoyant once again. There is a mention in 1910 of a total of 4,852 British-built threshing machines working in Argentina alone, of which no fewer than 1,926 were by Clayton & Shuttleworth.

The relatively peaceful late Victorian and Edwardian eras come to an abrupt end in 1914 at the outbreak of the Great War. Sinking of merchant shipping by German U-boats put greater pressure on the production of home-grown crops not just to feed the population but also to keep thousands of horses and mules in fodder.

It is difficult to imagine now, quite how many horses there used to be in Britain. In 1875, it was estimated that there were around 3.3 million horses in agricultural use alone. By 1900 this figure had dropped to just over 1 million but this was only part of a total for all horses engaged in both trade and agriculture totalling 2.6 million. Indeed the number of horses in use in London actually rose from 100;000 in 1850 to 300,000 in 1893. When the London Omnibus Company was founded in 1858, its fleet of 600 buses was drawn by an army of 6,000 horses. In one of Oscar Wilde's diaries, he described his overriding memory of London as being the smell of hay. At the beginning of World War One, the British Expeditionary Force took 25,000 horses with them to France. In the first 12 days of the War, the government purchased no less than an additional 168,000. By the time of the Armistice late in 1918, there were over 475,000 horses in use by the army and it has been estimated that the total number of British horses and mules in use for military purposes in 1918 exceeded one million.

During World War One, the amount of material required by this sea of livestock is just as difficult to imagine. Indeed no less than 5½ million tons of fodder was shipped from Britain for use on the western front. There was a vast market not only for grain, but also straw. As well as cut for chaff as fodder, it was also required for bedding. There was still a brisk trade to supply material for thatching, preferably in sheaf or bundle form for easy handling by pitchfork. From this, a profile of the output required from the threshing machine can be visualised. But at the end of the war, things were about to change.

By 1920, the number of heavy draught horses had fallen to 1.4 million and by the outbreak of World War Two, estimates for the number of horses working on the land vary from to 700 000 to 850 000. Even so, the tractor was still seen as very much secondary to the horse. In 1937, there were only the equivalent of 1.8 tractors for every 1000 acres, and even in the United States, the number of tractors did not exceed that of horses and mules until 1954. The low point for numbers of heavy draught horses in Britain would have come around 1960, and despite an upsurge in interest in shire horses, a survey conducted in 1986 could only identify 10,000, meaning a reduction of more than 300 times within a century.

After 1916, much pasture had been returned to arable use and the 1917 Corn Production Act was introduced so as to guarantee corn prices, exactly 100 years after its dubious predecessor. Interestingly, this legislation did not extend to barley. This was because of its use by the brewing industry, hence its exclusion following pressure from temperance lobby. When the war ended, the area under tillage fell from peak of 12.4m acres in 1918 to less than 10 m acres by end of 1920's. A second Corn Production Act in 1920 reaffirmed the 1917 Act, but in the following year, it was repealed, a decision called by farmers of the time "The Great Betrayal". Between Spring 1921 and Autumn 1922, not only did agricultural prices halve, but rental and labour costs rose to double pre-war levels. Compensation had been paid at the rate of £3 per acre for wheat and £4 per acre for oats, but between 1929 and 1933, the great depression caused farm prices to fall by 33%. By the end of 1933, at their lowest point, prices were at a level only marginally higher than the 1914 figure. The Agriculture Act of 1937 introduced deficiency payments for oats and barley and by the following year, some recovery had been made but agricultural prices were still only 33% higher than the 1914 figure. By 1928 world wheat production had risen to a figure 25% higher than 1914 level and by 1934, no less than 67% of all food in Britain was imported.

Just as in the first World War, the Second World War prompted the government to authorise grants to change from pasture to arable use. This was the target of the 1939 Agricultural Development Act which was supported by the Agricultural Research Council's "Plough-up" campaign. This recommended that 1,000,000 acres of pasture should be turned over to tillage but this was calculated to be equivalent to a yield increase of only 5%. To put his into context, improved farming methods saw British wheat production increase by a factor of 3 from 1939 to 2004.

Going back to the Armistice of 1918, four years of trench warfare had consumed an inordinate amount of wood resulting in a severe shortage of seasoned timber. This had been the preferred material for the framing of threshing machines, although some had been built with steel frames. Worse for the manufacturers was the repudiation of overseas debts. As one example, the new Bolshevik Russian government that seized power in 1917 refused to pay Garretts the £200,000 owed, an enormous sum at the time. As another example, Ransomes had sold 50 threshing sets to Turkey in early 1914 but never received payment for them. A number of builders of threshing equipment joined a consortium of 14 companies called the Agricultural and General Engineering Group or AGE. This managed to keep going during the 1920's but collapsed as result of the Great Depression in 1932.

Around this time, pneumatic tyres began to appear and tractors became more popular. Equipped with these new rubber tyres rather than the earlier spade-lugged steel wheels, road work became much more viable. Efficiency also improved as TVO gradually gave way to diesel. As a source of motive power that required no lengthy warming-up period, no boiler wash-outs, less routine maintenance, and one that could be left to run unattended for lengthy periods, this was a distinct improvement over the traction engine. If the high-point of thresher production was just before the First World War, the 1930's were lean years, not least because the quality of the machines produced 30 years before meant that they did not require replacement. A Royal Agricultural Society report of some years earlier had made the point that machines should exhibit "good workmanship and simple robust designs aimed at farmers, not engineers". A good example of this is a Ransomes machine bought in 1902. It was worked locally without a problem for over 55 years by 3 generations of the same family. World War Two saw an upturn in demand for home-grown produce but demand for straw as fodder or bedding for working horses was much lower than it had been 25 years before. American lend-lease tractors virtually put an end to the agricultural traction engine and common use of the threshing machine ended soon after with the arrival of the combine harvester. The earliest examples originated in the United States and Canada where wide-open spaces and a shortage of labour made them a very attractive option. Clayton and Shuttleworth were the first British manufacturer to produce a combine in 1928 but before World War Two, very few machines were in use in the UK. They were less well-suited to a country that had smaller fields, narrow gates, and a much damper climate. Apart from the large capital outlay involved, farmers were less happy with a machine that produced straw of much lower quality. Their mode of operation also meant that many weeds were recycled directly to grow again.

But requirements had changed. A vicious circle of fewer and fewer working horses compounded by less and less thatching meant that straw became virtually valueless. Selective propagation of cereals with the emphasis on more and more grain on shorter and shorter straw meant that combine harvesters could handle larger volumes of grain faster and faster, particularly if the cut height could be raised to further reduce the amount of straw passing through the machine. In terms of overall efficiency, this approach could be seen as wasteful, but the reduction in labour costs more than made up for the small amount of grain lost and the handling of "useless" straw. Developments in chemicals meant that nitrogen values could be increased and the resultant weakness in stalks could also be corrected. In addition to this, effective selective herbicides could cure the weed reseeding problem. The role for which the threshing machine had been designed to fulfil... had changed forever.

THRESHER PATENTS			
	1732	Menzies	Machine for thrashing grain.
13 February	1734	Menzies	Machine for thrashing grain.
21 June	1785	Winlaw	Mill for separating grain from straw.
9 April	1788	Meikle	Mill for separating grain from straw.
21 February	1792	Willoughby	Machine for thrashing corn.
19 February	1795	Jubb	Machine for thrashing and winnowing corn.
2 June	1795	Wigfull	Machine for separating grain from straw.
31 October	1796	Steedman	Thrashing machinery.
4 July	1797	Maule	Machine for cleaning the grain from straw.
5 June	1797	Palmer	Apparatus for cleaning the grain from straw.
9 November	1799	Tunstall	Machine for thrashing all kinds of grain.
6 December	1799	Palmer	Machine for cleaning grain/cutting straw.
19 June	1802	Lester	Machine for separating the grain from straw.
18 May	1804	Burrell	Thrashing machines.
30 October	1804	Noon	Thrashing machines with loose beaters.
16 January	1805	Lester	Engine for separating grain etc from straw.
5 February	1805	John Ball	Thrashing machines.
5 February	1805	Perkins	Thrashing machines.
23 November	1805	Lambert	Improved thrashing machinery and a windlass.
21 November	1807	Lester	Machine for separating corn from straw.
31 October	1808	Andrews	Thrashing machine.
23 January	1810	Cox	Thrashing machine.
22 May	1810	Onions	Thrashing machine.
29 June	1813	Todd	Separating the grain from straw.
27 September	1814	Lister	Machine for separating the grain from straw.
3 December	1817	Wild	Machine for separating corn from straw.

THE GREAT DIVIDE			
7 May	1840	Atkinson	Improved thrashing machine.
1 October	1840	Mackelcan	Improved thrashing machine.
21 January	1841	Cooper	Improved thrashing machine.
28 January	1841	Prior	Improved thrashing machine.
2 August	1842	Dry	Improvements in thrashing machines.

Chapter Twelve

EARLY ATTEMPTS 1

A Gaitin of Oats

Gaitin of oats.
a Band loosely tied. *b* to *c* Base of sheaf spread out.

Preamble

Before the arrival of the steam engine in the early 1800's, there were only 4 power sources available to the designers of threshing machines: man, the horse, the water wheel and to a lesser extent, the windmill. All of these sources suffer from the inherent disadvantage of low rotational speed. This can now be seen as an insurmountable handicap. Early on it was recognised that using a striking action to remove the ear from the stalk from a rotating bar required no small measure of outright speed. If the only available sources of power were inherently of such slow rotation, mechanical limits would be reached. Too great a size of step-up gearing would result in unacceptable frictional power losses, and ever larger drum diameters would make already large and bulky machines, totally unwieldy.

The compact steam engine was ideally suited for driving a high-speed threshing machine. Unfortunately its development coincided with a period of great unrest in the agricultural life of Britain. This is best demonstrated by the fact that despite no fewer than 24 patents relating to threshing by machine being registered in the 30 years up to 1815, there were NO such patents registered in the 23 years between 1817 and 1840. After this date, it is as if, after years of inactivity, engineers were able to turn their full attention to the design of the steam-driven thresher. This resulted in the emergence of a machine over the period of just a decade that was to remain in production for over a century changing only in relatively small detail until replaced by the less labour-intensive combine harvester.

Flails and Mills, Crushers and Wringers

As might be expected, the earliest attempts at the mechanisation of threshing were based on existing technology. Whilst some of the very first designs looked to the flour mill, most sought to replicate the action of the flail. Patent number 92A was taken out for such a machine, operated by cranks, by the Moravian knight, Sir John Christopher Van Berg in 1636, and records exist of a similar machine built by Hohlfreida in Saxony in 1711. Neither of these machines proved very successful, both showing a tendency for the hinged parts to self-destruct as speed increased. Even the famous Jethro Tull (1674 - 1741) is supposed to have designed a machine of this type, but lack of information suggests that he was no more successful than others pursuing similar designs.

A crude threshing machine was the subject of patent number 544 of 1732. Designed by Michael Menzies, an engineer-farmer from Dalkeith, East Lothian, it consisted of a series of flails attached to a rotating cylinder and was powered by a water wheel. It was capable of doing a considerable amount of work in a short time and attracted a good deal of attention. An article in an issue of the "Gentleman's Magazine" dating from 1752 suggested that the machine "gets 6% more grain than by thrashing with flails". Unfortunately frequent breakages pointed to the fact that a really successful machine was unlikely to be able to use the traditional action of the flail in its basis. A Mr Robert Brown wrote of the thresher "the flails are very soon destroyed from the velocity at which it was necessary to drive them". Menzies' machine was later reported as "failing to separate husks from grain in a satisfactory manner" and did not gain any wide popularity. However it DID stay together long enough to show that a mechanical method of threshing was possible, and Menzies took out a second patent in February 1734. He claimed that this later machine could deliver 1320 strokes per minute equivalent to 33 men "threshing briskly with flails", or more like 40 allowing for the men's rest periods. The Scottish Society of Improvers took an interest but were rather less optimistic. They rated the amount of work done by the one man managing the machine as being equivalent to that of 6 men with flails. An article in the 26 August edition of the "Caledonian Mercury" of 1735 quoted prices for threshing machines available at the time. One Andrew Good of College Wynd, Edinburgh, was offering machines which appear to be to Menzies' patent for £30 for the "4-man-equivalent" machine and £45 for the "6-man-equivalent" version. This was not seen as particularly good value for money. As a price comparison, the paper reckoned the cost of using the machines as equivalent to £7 10s per man, when a servant could be employed for a whole year for around £14.00.

An illustration of a French machine dating from 1765 shows two men hand-cranking an open-framed wooden drum fitted with around 30 wooden clubs arranged to strike along the length of a fixed board. Speed of rotation could only have been moderate since there does not appear to be any form of gearing.

Fig 261: French machine of 1765

Another illustration shows a machine built in Philadelphia in 1778 using a row of 6 three-pointed cranks mounted on a horizontal shaft to operate a total of 18 angle-shaped flails attached by short lengths of chain. Unlike the earlier French machine, a gearing mechanism is shown which would provide a higher rotational speed, probably at the expense of earlier self-destruction. Yet the machine shows more than a passing resemblance to the modern flail mower and were it built as solidly as the modern machine, perhaps the question should be asked as to whether this approach to threshing might have proved more successful.

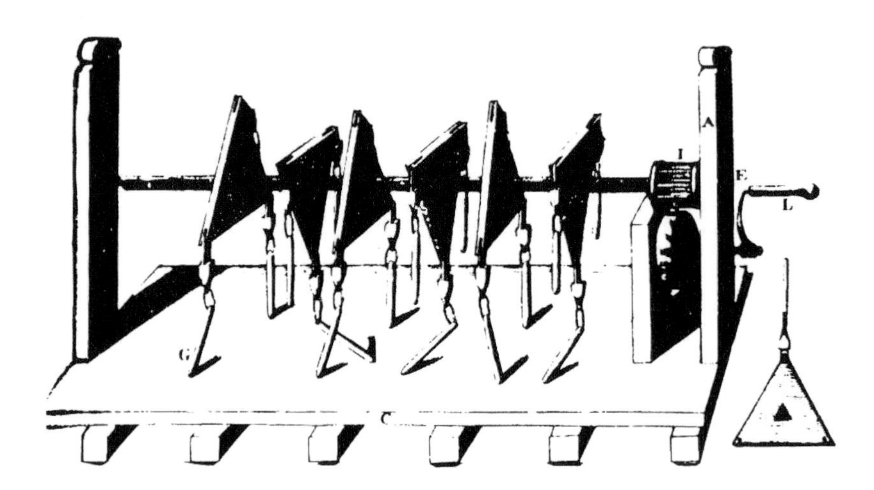

Fig 262: Philadelphia machine of 1778

In 1773, a Mr Oxley of Flodden, Northumberland erected a machine for Sir John Delaval. The machine had two feeding rollers, and instead of a drum, had a double set of arms connected by cross bars. The loose beater bars or "skutchers" were formed from pieces of wood about 3" wide and $1\frac{1}{2}$" thick at one end tapering to $\frac{3}{4}$" at the other. These were connected to the cross bars by leather straps. The unthreshed corn was laid on a board at the level of the centre of the skutchers and was fed in between a pair of fluted rollers. Once again, contemporary reports of the machine were not very favourable. It was said that very little corn could be passed at a time, and because the mechanism was exposed, "threshed material flew everywhere and clogged the machinery". One report went on to say "The operations of the machine were very defective" and it is not altogether surprising to read that "the machine was never made public, and I believe that Mr Oxley was very shy in showing it to strangers". Claims have been made that this machine was seen by Andrew Meikle but these have been strenuously denied.

In 1792 a Mr Willoughby of Bedford, Nottinghamshire, patented a machine consisting of 3 rows of 6 loose heavy wooden flails attached to a heavy wooden shaft. The machine was to be driven by a horse whim and the crop was hand-fed onto a grating through which the grain could fall.

Fig 263: Flail-type machine of 1792 (Willoughby)

In 1796 Mr M D Musigny in a letter to the Society of Arts, described another flail-type machine that he had seen. A set of 32 flails attached to a cylinder were driven by a single horse operating a whim of 40' diameter at a gearing of 20 to 1. Mr Musigny's opinion of the machine he described was not very high. The said "the unthrashed material was difficult to confine", as well as being "difficult to place and remove". He also went on to say that the corn "was thrown about in a most slovenly manner".

Also in the same year, John Steedman of Trentham, Stoke, Staffordshire, produced a very similar machine except that here, the crop was placed by hand on a rotating table where "a number of flails were made to play continually at a single spot" after which the crop was removed, again by hand.

Perhaps the design that replicated the action of the flail most exactly was a machine produced by one James Wardropp of Ampthill, Virginia in 1796 and an example of the machine was later brought to England. A horizontal shaft around which was tied a single spiral of thick rope operated rather like a camshaft lifting a row of 12 levers each in turn. This was known at the time rather charmingly as a "wallower" shaft. Each lever was attached by rope to a 12' long flat springy wooden blade, or as the contemporary account has it, "elastic rod". One end of each rod was securely attached to the frame of the machine; the other end was raised to a height of 3' 0" and then released as the camshaft turned. The spring of the blade itself brought it down sharply onto the crop that had been hand-fed onto a flat grid. The beating action of the blades released the grain that could then fall through the grid. The drive shaft was arranged so that it could be turned by two men.

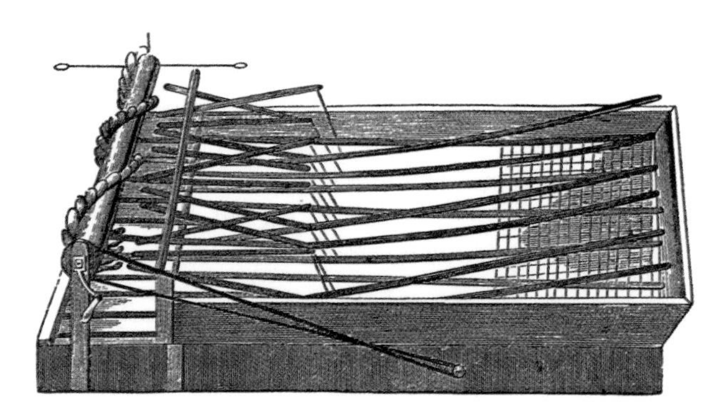

Fig 264: Flail-type machine of 1796 (Wardrop)

Another existing area of technology familiar to the pioneer designers was the flour mill. Once simple rotation was seen as the key to effective operation, a number of early designs featured a single vertical shaft with an overhead feed. One such design was patented by William Winlaw of Marylebone, London, in June 1785. His "conical rubbing mill" was the direct ancestor of the American axial thresher often referred to there as the "meat grinder", and in contemporary English accounts as the "coffee grinder". The large conical fluted cylinder "contained no loose parts" and rotated within a matching fluted vertical drum. Some reports of Winlaw's machine, which was powered by a water-mill, were fairly favourable but it is believed that only the ears were fed into the machine as opposed to the whole crop, and other accounts say that it had a tendency to grind some of the corn into flour.

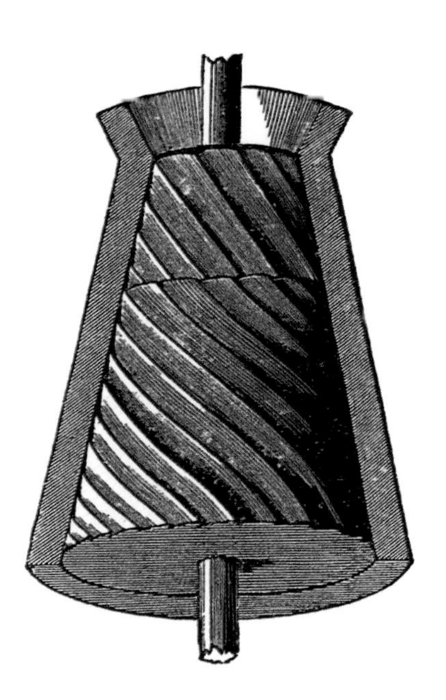

Fig 265: Mill-Type machine of 1785 (Winlaw)

A similar design by William Spencer Dix in 1797 drew similar criticism in that his "mill" damaged the grain and that it was difficult to feed. Here it would appear that a slight improvement had been made in that this later machine was attempting to handle the whole crop.

A third area of technology to be used as a basis was that of the ore crushing mill. One John Lloyd of Hereford demonstrated a machine to the Society of Arts in 1761 consisting of "a gate weighing around half a hundredweight, rising and falling in the manner of a sash window" to beat out the grain. It was fed by a hopper arrangement, but no record of its efficacy has survived.

In 1769, William Eves of Swillington, a small village just west of Leeds, designed a machine driven by a windmill where the crop was laid on a slowly-rotating table to be beaten by a series of falling bars rather in the manner of a set of Cornish tin stamps. A row of 15 heavy cast-iron bars with the tops turned over, were lifted, one at a time, by a large diameter drum fitted with four diagonal rows of pegs acting like a giant camshaft. The illustration of the gearing mechanism suggests that the one revolution of the table would be equivalent to around 40 turns of the drum resulting in some 2400 impacts.

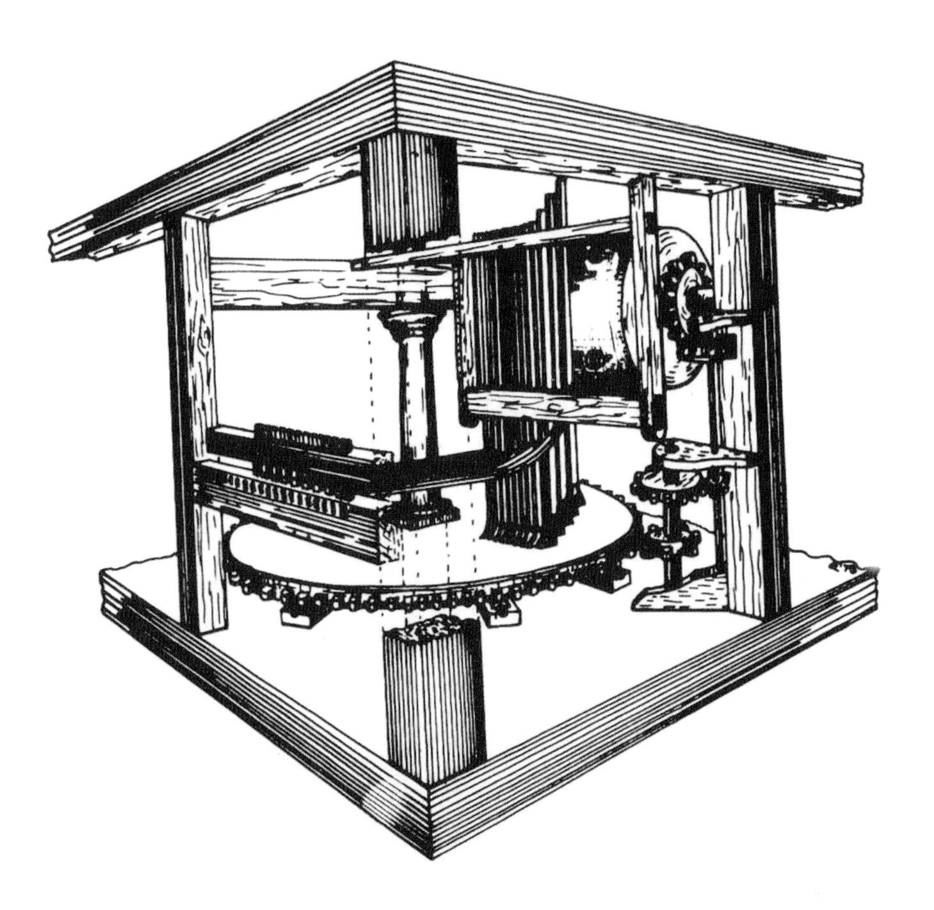

Fig 266: Crusher-type machine of 1769 (Eves)

In 1772, a Mr Ilderton, a farmer near Alnwick, Northumberland, working with a Mr Stuart, used a slightly different approach. Instead of relying on pure impact to dislodge the grain, a measure of pressing or rubbing out the corn was introduced by using what some have termed a "wringer" type machine. This employed a number of cylinders. There were two principal rollers each of 6' diameter. One was variously described as "fluted" or "corrugated", and the other "cellular" or having an "indented drum". These two revolved between a series of smaller corrugated rollers. The friction between the corrugations of the rollers and the crop was depended upon to squeeze the grain from the heads. The smaller cylinders were held against the larger cylinder by "stout springs" and it was claimed that the tension of these springs could be varied to suit different crop conditions. This machine was experimented with for some time but was found to be impractical, being slow in operation and liable to crack the grain. Contemporary reports said that threshing was "in several instances imperfectly performed". However the principle of incorporating a method of adjustment to suit varying crop conditions was an important step forward. As will be seen, one of the key features of the modern thresher is its range of adjustment making it able to deal with a whole variety of material to be processed.

On a similar basis, in 1795, a Mr Jubb of Lewes, East Sussex patented a simpler machine consisting of a pair of meshing 4-pointed solid wooden rollers set one above the other. The crop was fed "heads-first" from a flat loading board through a pair of corrugated rollers. As the crop was held by the feed rollers, a certain amount of beating action was received from edges of the main rollers. This was followed by a squeezing action between the main rollers. The grain fell into a hopper which discharged into a rectangular duct through which passed the blast from a fan. This fan was of larger diameter than the main rollers but of the same width and was belt-driven from them at around 1¾ drum speed. Schematically, this layout is very similar to that of a modern machine.

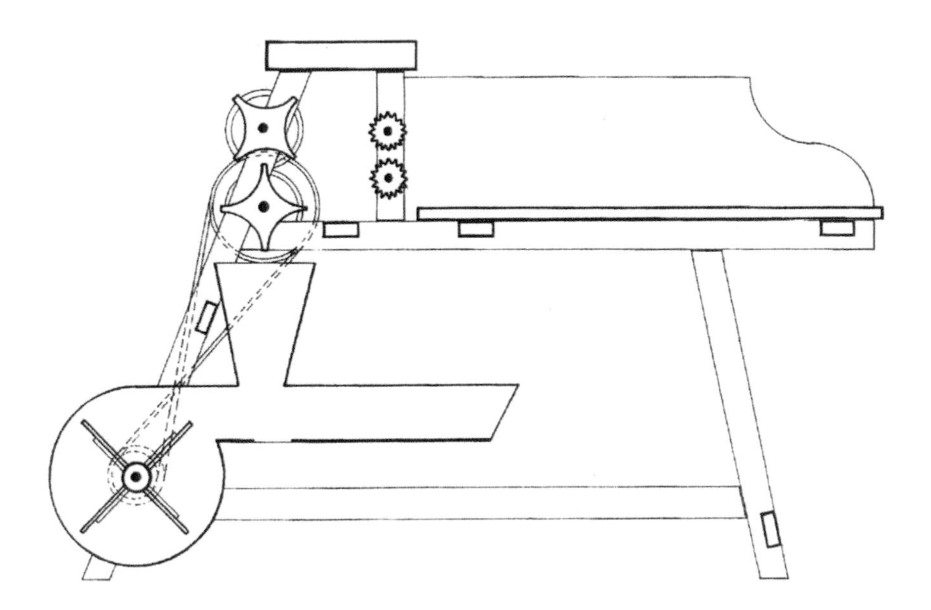

Fig 267: Wringer-type machine of 1795 (Jubb)

The relatively few contemporary illustrations available give little insight as to the scale of these early machines. Closer inspection of the few drawings that do exist reveals a dramatic polarisation in terms of size. The smaller hand-operated machines look to be around the size of a chest-of-drawers, whereas the larger pieces of barn machinery would fill the entire upper floor of a modern house. None of these early machines could be said to have worked really successfully. Despite many attempts with alternative systems, a horizontal rotating drum fitted with rigidly-attached beater bars was to become the universally-adopted system.

Chapter Thirteen

EARLY ATTEMPTS 2

At work in the barn

In 1753, Michael Stirling of Dunblane, Perthshire, devised a machine based on the principle of the flax-scutching mills of the period. Before flax fibres can be spun into linen, they must be separated from the rest of the stalk. To remove coarse fibres the flax is "broken", the straw being broken up into small, short pieces called shives, while leaving the fibre itself unharmed. The material is then "scutched". The straw is scraped away from the fibre in a machine that beats the fibres with a number of whirling paddles or blades. The fibres are then drawn through "heckles", which act like combs and which straighten and separate the long fibres.

Stirling's machine used a vertical cylinder 8' 0" diameter by 3' 6" high within which rotated a shaft at "considerable velocity". Four cross arms were mounted on the shaft but this time, everything was mounted completely rigidly. This proved to be a very significant change and is believed that Stirling's machine was the first to operate using this principle of all fixed components. Sheaves were fed in at the top of the cylinder and grain and straw were pressed through an opening in the floor to be separated manually using "rakes and fanners". Reports on the performance of the machine were less than enthusiastic claiming that it "breaks off the ears of barley and wheat instead of clearing them of the grain" and that "at best it is only fit for oats". Despite these criticisms, there are records that show that his machine was used by himself for several years, and was even imitated by others. As with Winlaw's axial thresher, it is thought likely that only the heads of the crop were fed into the machine as opposed to the whole crop.

In 1758, another Scottish farmer named Leckie succeeded in improving upon the Stirling machine. Using the same basic arrangement, the crop was fed in above the vertical cylinder, the rapidly-revolving arms beating the grain out of the straw during its downward passage. Both grain and chaff fell in a pile at the bottom. This approach correctly applied the basic principle behind flail threshing, rather than the action itself, and finally demonstrated the superiority of simple rotary motion without loose components.

Around 1785 on one of his tours through Northumberland, Sir Francis Kinloch of Gilmerton, west of Perth, happened to see the machine built for Mr Alderton by a Scotsman named John Raistrick. Kinloch commissioned a small replica which was built by a Mr Smart of Morpeth, Northumberland. He then arranged for number of improvements to be incorporated. The revised model was built by the 16 year old William Howden who went on to set up the Phoenix foundry at Boston in 1803 to produce threshers and mill machinery and later, a number of portable engines bearing his own name. As Sir Francis Kinloch had no suitable power source of his own for the new machine, he sent it to one Andrew Meikle, an engineer of Houston Mill near Tyningham, near Dunbar, East Lothian.

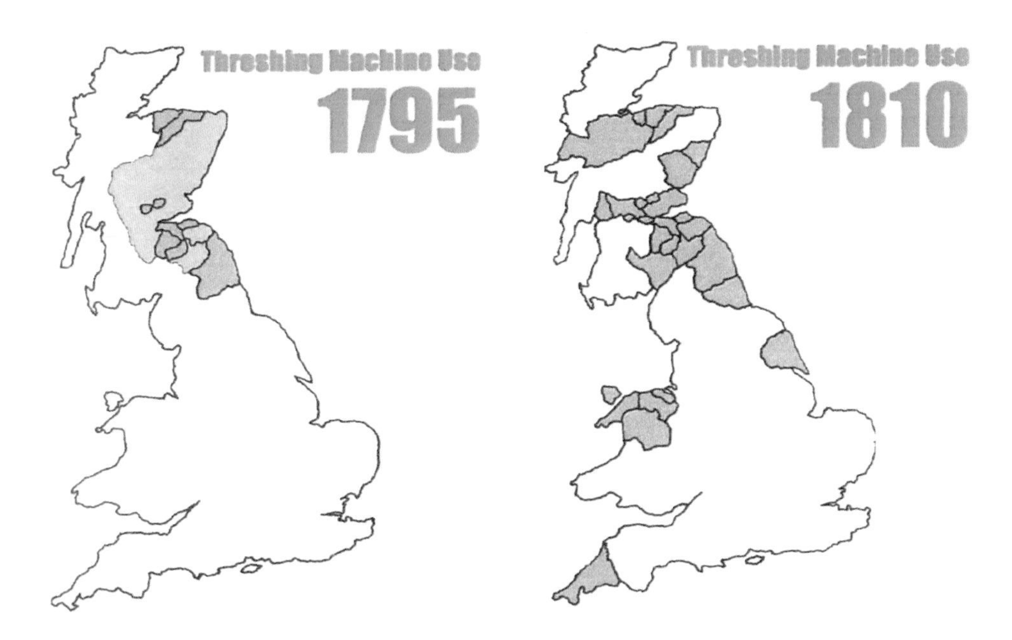

Fig 268: Map showing use of threshers in 1795

Fig 269: Map showing use of threshers in 1810

Here it could be tested using the water-wheel of Mr Meikle's barley mill. Unfortunately during the trial, the machine was "torn to pieces". Undaunted, some years later Sir Francis commissioned a similar full-size machine based on the model sent to Mr Meikle but unfortunately, this was to meet the same fate as its smaller predecessor.

Andrew Meikle (1719 - 1811) is generally credited as being the inventor of the first practical threshing machine. A portrait of the Scotsman by one A Reddock hangs in the National Portrait Gallery. Andrew produced two early unsuccessful designs in 1778, then in 1785, he constructed a third thresher which embodied many of the essential features of the modern machine. This machine, like all its predecessors, was a thresher only and not the combined thresher and winnower as generally thought of today. The 1785 design thresher is usually described as an improvement on a "Northumberland Design" that in turn had been based on principles used on contemporary flax-scutching machines". The Northumberland Design referred to is almost certainly the model sent to Meikle by Sir Francis Kinloch mentioned above. The performance of this earlier machine was greatly improved by the fitting of hard oak pegs to the 4 beater bars, and later, by the fitting of similar pegs to the concave.

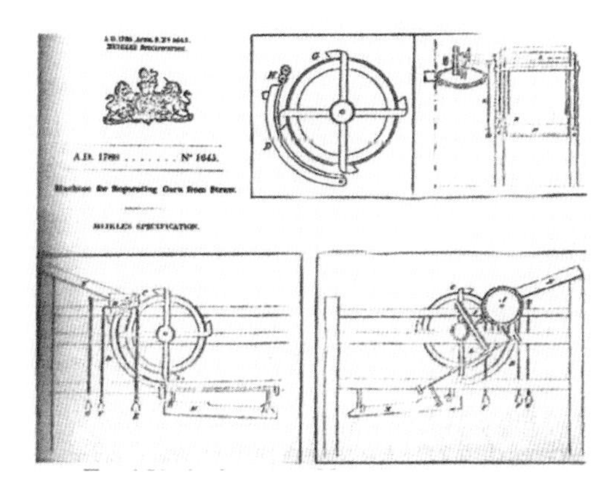

Fig 270: Andrew Meikle's Patent of 1788

Perhaps the key to the success of Meikle's later design was the outright speed of the machine. He is quoted as saying that "no machine is capable of doing the work in a perfect manner, that does not move with a velocity exceeding 2,000 feet of the circumference of the drum in a minute". This compares with the 6,000 ft/min for a modern machine. He constructed a strong framed drum with fixed bars that beat rather than rubbed the grain. Meikle's drum revolved at 200 to 250 revolutions per minute, as against the 1000 to 1100 rpm of a modern machine. The drum cleared the solid concave or as he called it, "breasting", by some 5/8". The loosened sheaves were fed, ears first, from a feeding board between a pair of fluted rollers that held the crop against the drum. This drum was fitted with 4 wooden cross members, sheathed in iron, parallel to its axle, and these struck the ears of corn as they protruded from the rollers knocking out the grain. He modified the profile of the original square-edged bars, undercutting them at an angle of about 45 degrees to give them a better angle of attack. Some years later, John Morton, manager of Whitfield experimental farm, Gloucestershire, reduced this angle still further to more like 30 degrees.

Another feature of the Meikle machine that was to be adopted almost universally was the shaker or walker mechanism. This simple reciprocating single tray or "jogging screen" later evolved into the series of parallel bars known in Britain as "harps" and in North America as "vibrators".

Andrew Meikle, helped by his son George, built the first of these machines for a Mr Stein of Kilbogie in 1787 on the understanding that if failed to operate effectively, no cost would be incurred. Meikle's second thresher was supplied to a Mr Selby of Middleton, Northumberland, and was powered by a windmill. The performance of previous machines using this method had been poor because of the constantly changing wind speed. Meikle solved this problem by patenting a sail furling mechanism that kept the rotational speed of the thresher more even. This allowed the thresher to perform very successfully and its design was copied by a number of local mill-wrights.

A Meikle machine was taken to New York by Baron Pollnitz in 1788 where it successfully threshed 70 bushels in a day when operated by one man and a boy. This was seen as being around 10 times the amount that they could thresh using flails. The first American patent relating to threshing was granted just 3 years later to Samuel Mulliken of Philadelphia in 1791, and a machine demonstrated in 1796 by no less a person than Thomas Jefferson was said to be capable of threshing 120 bushels a day.

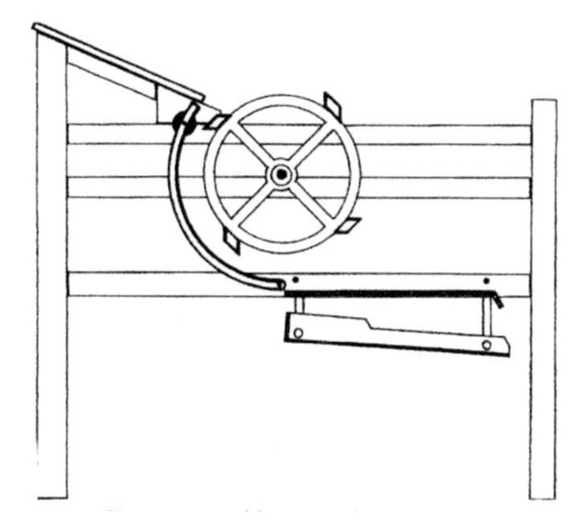

Fig 271: Meikle's Machine of 1790

But Meikle did not attempt to file a patent for his idea until 1788. When he eventually sought to do so, he encountered some difficulty because others were already seen to be producing similar machines. It was claimed that a thresher had been built in Gothenberg, Sweden which pre-dated Meikle's machine. In fact a pipe-layer, one Andrew Blackwood had been working in the area and had seen a need for such a machine. He contacted a fellow tradesman, John Girvan, who he knew had assisted at the building of Mr Stein's machine and this was how Sweden's first thresher was born. John Raistrick, who had built the unsuccessful machine for Mr Alderton, also claimed to have built an improved version very much on the lines of Meikle's later thresher before the patent had been filed. There are records of two such machine being built, one in Surrey and one in Oxfordshire. But this was not until 1790 so the general consensus of opinion is that Meikle's was the first completely successful threshing machine as shown in patent number 1645. Mr Raistrick is also mentioned by Thomas Stone in Young's Agricultural Survey of 1800 as selling equipment to farmers at inflated prices. The article levels the criticism that the savings over traditional methods offered by Raistrick could not be substantiated. This very reminiscent of the article that appeared in the "Caledonian Mercury" of some 65 years earlier.

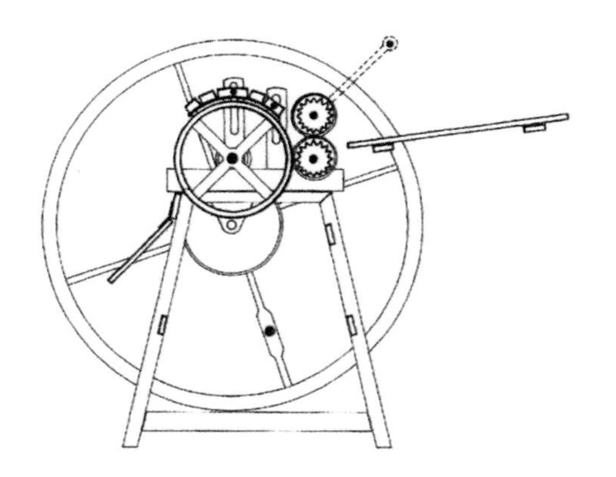

Fig 272: Machine with solid overshot concave - 1793

Meikle's problems with the patent led a group of gentlemen headed by Sir John Sinclair to raise a fund to "render Mr Meikle's declining years more comfortable" which goes to show how little he had benefited financially from his invention.

Several other people also claimed to have produced successful machines before Meikle. A Mr Gladstone from Castle Douglas,15 miles south-west of Dumfries is reported to have demonstrated a machine in 1794. Jeremiah Bailey of Chillingham, 12 miles north-west of Alnwick, used a machine driven by a water-wheel that was reported as being capable of threshing 35 bushels of wheat in an hour. He is also credited as being the first to fit a winnower or fanner directly below the drum. The long list of claimants also includes a Mr Palmer in 1799.

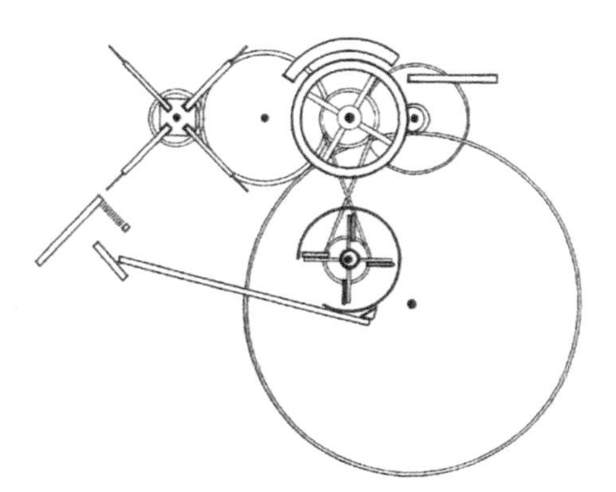

Fig 273: Machine with solid overshot concave - 1798

What became known as the Scotch Threshing Machine was seen comparatively rarely in England. A reference in 1852 states that these are only found "at the homes of Noblemen such as Holkham Hall, on the north Norfolk coast, and Whitfield, Gloucestershire, of which more later". These used the same corrugated roller feed and solid concave but here the concave was fitted above the drum. To use the parlance generally associated with water-wheels, this made the machine "overshot", rather than "undershot" as originally designed by Meikle which formed the basis for all future designs other than some of these early Scotch machines. Thus the grain fell away from the solid concave through the "skeleton" drum. It was said that whereas in a "Scotch" pattern machine, the grain removal was achieved entirely by beating, an "English" pattern machine also relied on a measure of rubbing action.

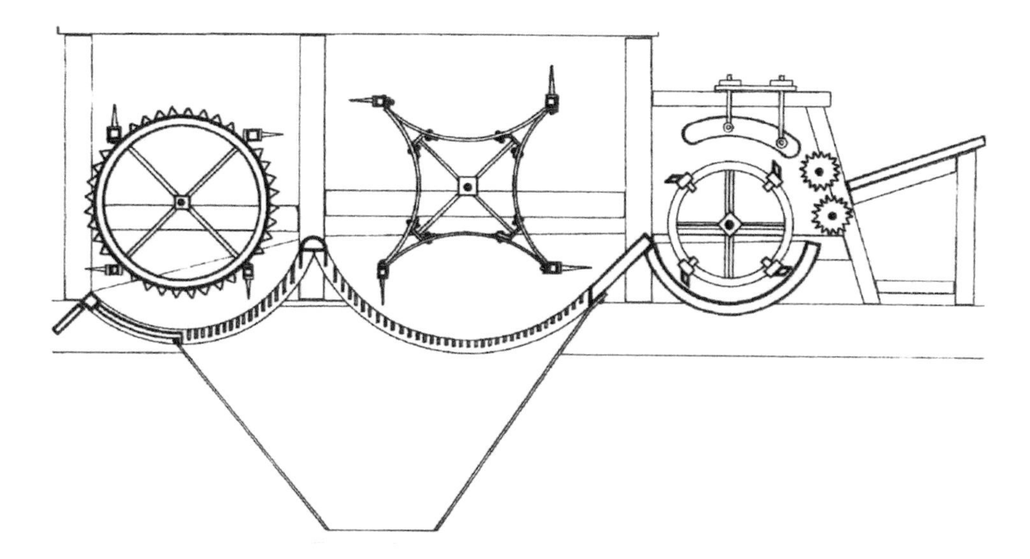

Fig 274: Section of a typical Scotch Thresher

The drum speed was increased to 350 rpm which is still less than a third of that of a modern machine. However a relatively small increase in drum diameter to 3' 6" gave a speed at the drum circumference of nearly 4000 ft/min which represents nearer two thirds of that of a modern machine. In fact the diameter could be anything between 2'6" and 5' 0". By way of comparison, the most popular size for a modern machine would be just 22". The feed was through a pair of cast-iron rollers of 4" diameter and the better machines used a fluted pattern that was much less prone to clogging, especially when the crop to be fed was wet. Another revolving drum or rake took the straw from the threshing drum and passed it to a third which tossed the straw out of the machine. The straw thus passed over 3 drums where the "thorough rubbing and tossing" separated the grain and chaff from it. The feed by the corrugated rollers was geared so that the heads received 19 blows from the beater bars equivalent to almost 5 revolutions of the drum. The action of these secondary drums is very similar to that of more recent "back-acting" hay tedders, and represents another example of haymaking inheriting cereal production technology, rather in the same way as the mower was derived from the reaper.

Fig 275: Drawing of an 1880's Hay Tedder

Fig 276: Photograph of a 1950's Hay Tedder

A large sieve extended below all 3 drums with the grain that fell through the sieve passing via a hopper to the separate fanning mills or winnowers located on the floor below. These blew away any remaining chaff. All these machines could be arranged so that they could be powered via a single set of pulleys. Andrew Meikle had filed a joint patent with his son George in 1768 for a "double blast cleaning machine" where the blast from the first fan removed the chaff and the blast from a second fan produced the final "finishing".

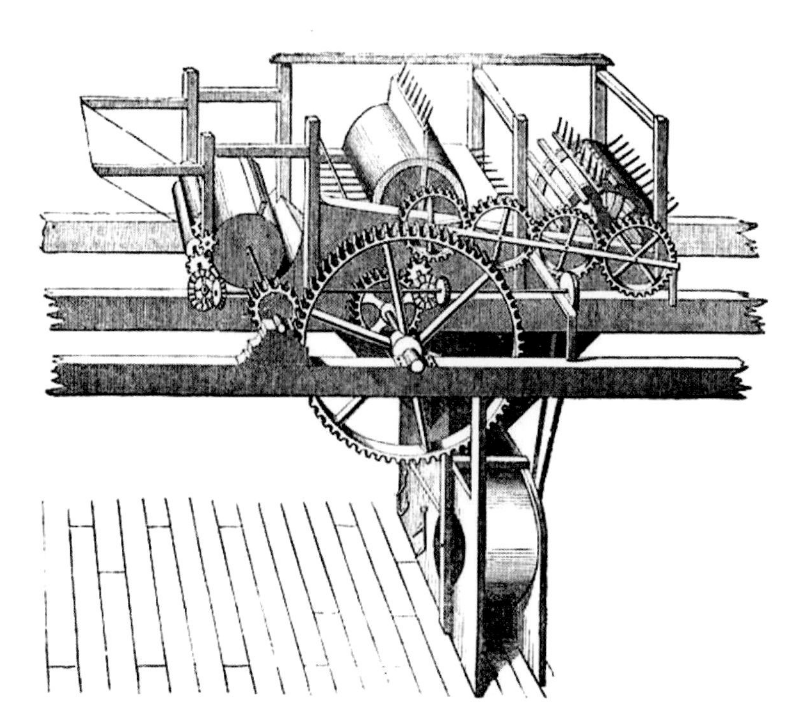

Fig 277: Contemporary Drawing of a 1790's Scotch Thresher

A distinctive feature of many of these early Scotch threshers was the fitting of oak pegs to the drums as first used on Meikle's early designs. After 1790 or so, the trend was to fit the concave with matching oak pegs which squeezed the straw further to release the grain. The oak pegs were later replaced with pegs of iron and this became the Scottish Peg-mill.

This design was to find favour on the other side of the Atlantic and the first American patent for a peg-type machine was granted to a Mr S Turner of New York. The earliest machines, on which Turner's design was based, required 4 men to operate them; two to turn the handles attached to the drum, one to feed, and one to rake away the straw. Still in America, a Mr Howe devised a machine driven by a horse walking on a rotating table that drove a vertical drum thus making this an axial thresher. The drum itself was a solid cylinder with pegs operating within a solid concave cylinder so that the threshed material fell from the bottom. Another early patent was taken out by a Mr A Savage in 1822 and an improved design was patented by Mr A Douglass in 1826.

From these originated the famous American "Groundhog" thresher so named, not just because of its shape, but also because this lightweight compact machine needed to be firmly staked to the ground to prevent it moving about when in operation. It could thresh up to 100 bushels of wheat in a day. Joseph Hall of Rochester, New York, built his first threshers in the early 1830's using a solid wooden drum turned on a lathe and fitted with iron spikes. The roaring noise made by these machines in operation resulted in their nickname of "Bull" threshers. Jacob Wemple of Mineyville, New York, introduced an improved tooth design to existing threshers that prevented them from coming loose, or worse still becoming detached. In 1840, he joined George Westinghouse, later of air-brake fame, to produce his own machines. Similarly, Jerome Increase Case modified existing Groundhog threshers before building machines of his own in his hometown of Williamstown, New York. This was before he opened his famous factory in Racine, Wisconsin, in 1847.

Although nowhere near as popular in England, a patent was granted to Joseph Atkinson of Bramham Hall, Yorkshire in 1840. This patent shows the peg-lined concave as covering 50% of the drum circumference. As late as 1855, an illustration, from the Cyclopaedia of Agriculture, shows a fixed threshing machine described as being 'on the

Scotch principle' which meant one based on one of the Meikle designs. Even in 1880, one writer noted that "such machines can still be found in nearly every farmstead in Scotland as a fixed thresher in the barn, though in many cases unused since the advent of the portable thresher". This is not altogether surprising; the sheer size of these early peg-mills poses a serious challengo anyone thinking of dismantling one.

Back in England, in 1795, a Mr Wigfull of Lynn, took out a patent with a view of combining the merits of the flail and the rotary machine. The crop was fed by a pair of corrugated rollers towards a rotating drum fitted with a series of beaters attached by short lengths of chain. The grain removed was then carried via a "shaking screen" and a canvas belt variously known as a "rolling cloth" or an "endless web" to receive the blast of a fan in order to perform the winnowing. The performance of similar earlier machines would suggest that self-destruction would be a distinct possibility, but the integral dressing arrangements have led some to see this as one of the first examples of the modern "combined" thresher. The first example was built for messrs Ede & Nicholls of Elm, near Wisbech, Cambridgeshire, and received praise in the "Repertory of Arts". It cost 100 gns and was originally designed to be driven by 4 horses; this was later increased to 6. Its daily output was rated at 15 to 20 quarters of wheat or 25 to 35 quarters of oats. Describing this machine in 1843, James Allen Ransome commented that the efficiency of these machines had increased by a factor of 4 in the intervening 50 years, and by then required only 2 or 3 horses to drive them.

Drawings exist for a modern-style drum and concave in a design by H P Lee of Maidenhead in 1802, but it is uncertain if that machine was ever built. A later machine of 1810 submitted to the Society of Arts, received a Gold Medal. It featured a 4 vane drum operating within a concave and was describe by J A Ransome in 1843 as "the model upon which many of the machines in England continue to be made". Some even cite this as the first machine specifically designed to operate without feed rollers, but this is perhaps an exaggeration.

William Lester of Paddington, London, patented a number of thresher designs between 1802 and 1805 all of which were claimed to avoid the scutching principle in order to reduce the amount of broken straw. Mr Lester is said to have been the first man to describe himself as an agricultural engineer, and was reported as the first to have built special traction engines expressly designed to pull and to drive his threshing machines as early as 1814. It was not until after 1800 that threshers were being produced in any numbers that incorporated the winnowing operation within an extended Meikle-type peg-drum design so making it possible to thresh, clean, and deliver grain in one operation for the first time.

The 1881 Household Cyclopedia said of these machines: "By the addition of rakes, or shakers, and two pairs of fanners, all driven by the same machinery, the different processes of thrashing, shaking, and winnowing are now all at once performed, and the grain immediately prepared for the public market. When it is added, that the quantity of grain gained from the superior powers of the machine is fully equal to a twentieth part of the crop, and that, in some cases, the expense of thrashing and cleaning the grain is considerably less than what was formerly paid for cleaning it alone, the immense saving arising from the invention will at once be seen".

In 1804, Joseph Burrell filed Patent no 2757 of 16 June for an "Improved Threshing Machine". It was arranged so as to make it possible to thresh corn once, twice, or three times. This was claimed to achieve better separation of the grain from the straw than would be possible with other machines of the time which relied on a single pass. The design featured 3 pairs of revolving fluted rollers in addition to the revolving drum with beaters. The operation of the machine involved the crop passing through the first set of rollers to the drum, back through the second set of rollers to the drum again, and similarly for a third time. This principle was adopted on many later machines. However some urged caution fearing that passing barley through a machine more than once could "injure samples intended for malting".In the same year, Ransomes, Sims, and Jefferies of Ipswich, Suffolk, built the first of a series of threshers to a design which they continued to produce for almost 40 years. The original carried a drum of 19½" diameter with a width of 42". This drum was built as a wooden barrel on metal rings. The surface was completely covered with cast-iron plates carrying ribs projecting ¼" in a spiral pattern resulting in the whole surface being divided into a pattern of squares and lozenges. The concave was built from a number of wooden battens and covered about a quarter of the circumference of the drum. It was mounted above it on chains so that it could move away from the drum in the event of a blockage. Each of the bearings for the mainshaft were described as being "of plain wood strengthened by a small piece of bar iron".

A major step forward...

A Devonshire mechanic named John Ball (or Balls) who had moved to Letheringsett north-east of Fakenham, Norfolk, was the designer of the first effective threshing machine to use the principle of passing the grains between a drum with beater bars, against an open concave. The wooden drum was fitted with 2 wrought-iron bars. The concave covered about one third of the drum and was framed with bars threaded with closely-spaced wires. On Ball's original machine, the bars were made of wood but this was soon changed to iron. This "English patent" or "open" type machine quickly formed the basis of future technological developments. This was not least because much less care was required when feeding grain directly into the cylinder/concave surfaces of the English design when compared with a machine using a corrugated roller feed. As with virtually all machines of the time, the crop was fed endwise, heads first, and the chaff-and-grain mixture was separated by winnowing. Significantly, the omission of the fluted feed cylinders from this design would allow the crop to be fed crosswise, providing that the drum was made wide enough. This would lead to the next major step forward in design.

All early threshers had the crop fed "ends-first", usually "heads-first", and usually via a pair of rollers. It was said that these machines could be fed more evenly if the stalks were all of the same length. This was easy if the crop had been cut using a sickle, but far less so if the more popular and efficient scythe had been used. The Scotch pattern threshers with their peg drums were exclusively fed "ends-first". This design was taken up in North America, and when the first combine harvesters appeared, they retained the same system so that they could never deliver the quality st raw sample possible with an "English" pattern machine.

The difference in action between English and Scotch machines may be seen from the following data:

Detail	Scotch	English
Proportion of Drum covered by Concave	30%	75%
Clearance between Drum and Concave	3"	1½"
Maximum Beater Bar speed	3000 ft/min - 34 mph	4000 ft/min - 46 mph
Bushels of Wheat threshed per hour *	26	36

* These last figures were obtained by J A Ransome using machines driven by identical 4-horse whims. Ransome also added that the English pattern produced better quality straw and that many of the Scotch designs featured a "seconds" elevator to process the entire crop another time.

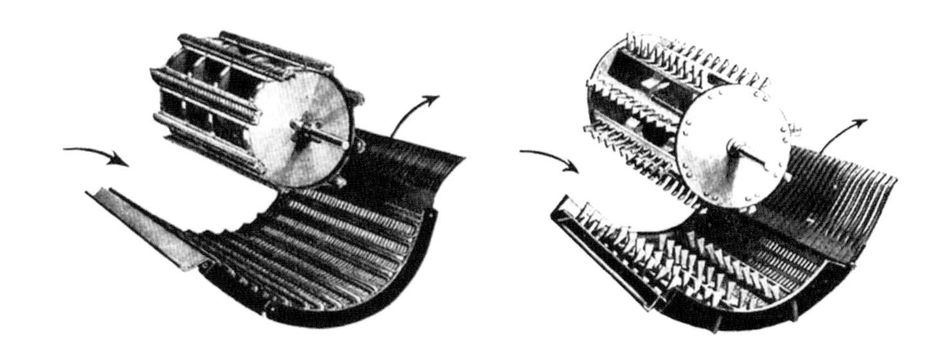

Fig 278: English Plain pattern Threshing Drum

Fig 279: Scotch / American Peg pattern Threshing Drum

John Ball's original machine was first demonstrated at the 1805 Great Holkham Sheep Shearing Show where it was driven by a horse-worked capstan. John Ball's daughter Sarah had married Richard Garrett shortly before, and production of the Ball design thresher was begun by the Garrett company in 1806. Burrells had also exhibited a thresher at the same Holkham Show described as "being capable of threshing corn in 3 stages in a single process to produce straw-free grain".

As may be seen from the above, the early development of the thresher was very gradual. If Meikle's original argument was right, namely that that improved performance could only be achieved by greater beater-bar speed, then no substantial improvement could be made until another form of motive power could be found. What was needed was a source of power with high rotational speed and this finally arrived in the form of the steam engine.

John Morton, manager of Lord Ducie's experimental farm of Whitfield, Gloucestershire, built a barn thresher around 1840 that drew the attention of Mr Ransome in his 1843 survey. Powered by a 6 HP steam engine, it was fed by a bucket elevator from below. The sheaves passed directly to the drum down an angled fixed feed board from the buckets without any feed rollers. The plain cylinder was of 18" diameter built from iron plate on a cast iron skeleton. The beater bars were arranged in a V formation facing in the direction of rotation of the drum. Each bar was constructed from angle iron and projected 7/8". The concave covered one third of the circumference of the drum and was built from 4 or 5 arched sections each 3" wide and jointed together. Across these ran square-section cast iron bars held in place by a number of bolts, each with a coil spring. This was intended to allow the concave to "give" in the event of material becoming jammed. This would have been very useful in view of the fact that the working clearance was limited to just 1/8"! Once again, the "open" design of concave was used so as to allow most of the grain to pass through it.

Around the same time, more attention started to be paid to the profile of beater bars, or as they might have been termed at the time, rubbers or scutchers. The number at this time was usually 4 or 5 and the traditional design was square-section wooden spars faced with iron plates. Some makers began to develop all-metal bars. Some used a circular iron rod, and others used a semi-circular pattern mounted so as to strike with the sharp edge. The next, and more important, development was to include spaces for grain along the length of the bar. These took on a variety of profiles, serrated, notched, or toothed.

Much emphasis has been placed on the importance of the threshing action of early machines, but another significant change was the incorporation of the various follow-up processes within the single unit. Before this, a separate fanner / duster would take grain fed from a hopper in the barn above. A feed roller was arranged to allow the grain to fall in a thin sheet in front of the fan blast. The heavy material landed on a sieve which directed it to the spout. Light material was blown away and small items fell through the sieve.

As has been mentioned previously, one of the first recorded uses of this type of winnowing machine in Britain was by the Laird of Saltoun, Andrew Fletcher. These were based on designs for barley mills and fanners that he had seen on his visits to Holland. These are believed to have originated in the Dutch East Indies where the Chinese had used similar machines to process rice. According to a series of letters by Robert Somerville of Haddington, the Laird sent Andrew Meikle's father James to Holland "so that he might learn the perfect art of 'sheeing' barley". When he returned, Meikle senior built a number of machines and took out a patent in 1710. Unfortunately he remained bound to the Laird, by an agreement that he should not profit personally from the design. It is perhaps no coincidence that James Meikle christened his son Andrew who went on to be the millwright at Houston Mill near Dunbar. Apart from these fanners at Saltoun barley mills, little is heard about these machines until around 1733 when Andrew Rogers of Cavers, Roxburghshire built his own fanning mill. The Meikle family remained involved with winnowing machines and in 1768, a patent in the names of A and R Meikle was filed for an improved design. At this time, it was common practice to pass the material through the fanner 2 or 3 times. Even so, this was still much faster than by hand. and such machines were reckoned to take "a few minutes per bushel" as opposed to around half an hour manually. It wasn't until 1798 that Mr R Douglas in his "Agricultural Survey of Roxburghshire" describes a fanner designed by a Mr Moodie of Lilliesheaf, south of Galashiels in the Scottish Borders, which was expected to complete the process at a single pass every time.

The early open designs were replaced by the fanning mill where the operation all took place in an enclosed box, a system of manually-operated blades or vanes creating air currents to blow away the chaff. Performance improved with the introduction of "scroll" pattern fan casings, patented by a Mr J Gooch of Northington, near Stroud in 1800, whose later design won a prize at the RASE's 1841 Liverpool Show. Another improved design of winnower was built by John Elmley in 1812. The performance of Gooch's scroll pattern fans was greatly improved by the addition of backward-curved blades designed by Mr R Clyburn from Uley, near Stroud, Gloucestershire, while working for Lord Ducie on his nearby Whitfield estate.

As well as improvements in winnowing, this period saw greater emphasis being placed on the whole area of finishing or dressing. In 1839 Mr T F Salter received a silver medal at the RASE's Cambridge Show for his "combined winnower and hummeller". Inside a fixed cylinder, some 6' 0" long and mounted at an angle, rotated a shaft fitted with a "spiral of 2" long blunt arms". Flow rate was controlled by 2 sliding shutters or "sluices", located one at either end of the cylinder. The grain was then discharged onto a series of reciprocating sieves, or as the contemporary account words it, "peculiar jerking motion by crank". It was also subjected to blasts of air provided by fans. It was claimed the one man and a boy could process 8 to 10 quarters of barley in a day. Messrs Hornsby of Spittlegate Ironworks, Grantham, incorporated a spiked roller working through a grating which was said to effectively separate chaff found in what is described as a "rough and pulsy state". It can be seen that the modern piler differs little from this design. Before these machines, barley would be laid across the barn floor in a 2" layer to be chopped all over by labourers using a special blunt knife known as a "hand-hummeller" or "barley-chopper". A minor improvement to this system consisted of a roller fitted with the same blunt blades.

A not dissimilar machine was produced by a Mr Grant of Granton, Aberdeenshire, to remove smut. This also used a 6' 0" long cylinder containing an upright shaft. This time the cylinder was constructed from wire mesh and instead of the blunt blades, the shaft carried a series of strong brushes. The grain was retained within the cylinder whilst the smut fell through the mesh. Interestingly, Mr Grant's smutter shows more than a passing resemblance to the modern rotary screen. More tellingly, J A Ransome comments in his 1843 book that such systems would not be economically viable "except for the very best quality output". This would be for human consumption rather than animal feed and a contemporary article states "After treatment, the grain assumes a bright, clean, polished appearance much appreciated by millers who know that it makes superior and brighter flour". Improvements were also being made in the area of straw handling. Following on from Meikle's simple single "jogging screen", Mr Docker based in Findon, a small village 5 miles south of Aberdeen took out a patent for an improved version in 1829. This was followed by an updated design that was patented by a Mr Rotchie of Melrose, 3 miles east of Galashiels, in 1837.

One machine fitted with these multiple "shaking harps" was that built by John C Morton, at Whitfield, referred to earlier. A contemporary account describes the machine as having "no fewer than 30 frames". Each of these were 6' 0" long by 2" deep and ¾" wide, with a ¾" gap between each frame. Within an overall width of 3' 9", all 30 operated in a reciprocating motion given by 2 crankshafts of 3½" stroke with alternate throws at 180 degrees. This was certainly the origin of the most popular design of shaker used on later machines. It was said that these mechanisms allowed a great quantity of "pulse" or "colder" to pass through, but that this could be removed when the shakers fed a vibrating trough or riddle, the forerunner of the modern cavings riddle. This was said "to effect a considerable saving in manual labour".

By 1820 there were a substantial number of fixed barn thrashing machines in the Northern England /Lowland Scotland area. By their very nature, all were farmer-owned. By contrast, the period between 1845 and 1850 saw a the biggest change with the introduction of portable thrashing sets, many, but not all, powered by portable steam engines. These were mainly in Eastern and Southern England with many of these being owned and operated by contractors. It is these machines that evolved into the modern thresher.

As early as 1811, Boston Agricultural Society had offered a prize of 20 gns to the first user of a "well-constructed portable thrashing machine capable of being easily removed from place to place, set at work without difficulty, and properly adapted to thrashing cole seed and corn." Here cole seed means rape seed. Mention is made of 3 local manufacturers of horse-whim driven threshers, these being James Coultas of Grantham (1811), Richard Gray of Boston (1810), and John Pinder of Grantham (1815). Here the term portable would probably have meant capable of being dismantled and loaded onto wagons, rather than a single unit mounted on wheels ready to be towed, The same term could also be applied to steam engines of the period. Looking back at these designs, a better description might be "semi-portable" or "demountable".

Despite the changes going on in machinery in the south and east, fixed barn threshers remained popular in the Scottish Borders region. The wheel was to come full circle with the very last of these fixed threshers. These were simply "modern" portable machines with their wheels removed, and mounted on stands.

Fig 280: A "modern" barn Thresher

During the above period, the cost of threshing changed dramatically. Although this was mainly due to mechanisation, at least part of the reason must have been because of increased labour costs, as may be seen from the figures below:

COSTS OF THRESHING PER QUARTER (8 BUSHELS)		
1794	2/2d	By hand
1811	5/6d - 7/0d	By hand
1850	3/6d	By hand
1850	1/7 ½d	By horse-driven thresher
1850	-/7 ½d	By steam-driven thresher

Ransomes, Sims, & Jefferies Limited
IPSWICH & LONDON

RULES FOR WORKING THRASHING MACHINES

1. OIL FOR BEARINGS
For ball bearings use only good quality grease free from both acid and alkali. Use only clean oil of good quality for other bearings and do not allow the working parts to heat.

2. FEEDING THE DRUM
Never begin feeding until the drum has attained its full speed. Feed regularly and spread the sheaf over the whole width of the drum.

3. KEEP THE MACHINE CLEAN
Clean the machine down, wipe all the oil off the outside of the bearings and cover it up every evening: keep the road wheels and axles well-greased.

4. KNOCKING OR UNUSUAL SOUNDS
Whenever a bolt nut or screw gets loose or shakes tighten it immediately.

5. ADJUSTMENT OF CONCAVE
Adjust the concave by the screws and nut at each side of the machine so that the grain is neither broken nor in left the straw.

6. REGULATING THE BLAST
The blast from the fan must be regulated by the wind boards so that the chaff is well separated, without blowing any kernels over the chaff board.

7. CHOBS IN THE FINISHED SAMPLE
If many chobs appear in the finished sample adjust the beaters of the chob cleaner above the sieve at the back of the machine (see directions below).

8. STRAPS
Put the straps as shown in the illustrations on each side of the machine. Take care to keep them tight especially the elevator bucket strap which is readily tightened by the buckles adjustable top bearings.

9. CAVINGS RIDDLE
The cavings riddle is made in two parts an extra part being also supplied with each machine. For wheat and rye both the parts with the small holes should be used. For barley, oats, peas, or beans, the extra front part with the large holes should be used.
Extra large beans require a special riddle.

Chapter Fourteen

DEVELOPMENTS AFTER 1840

Thresher Nameplate (C & S)

As far as threshing machine design is concerned, relatively little progress was made in the period immediately following the Napoleonic War. The widespread social unrest often resulted in threshers being deliberately targeted by angry mobs, and widespread improvements to these machines did not arrive until late in the 1840's. The economic climate remained uncertain and this era is sometimes referred to as "The Hungry Forties". Although the majority of machines in use in Scotland and the north of England were of the fixed Scotch type, those of the southern, and more particularly eastern counties of England were of portable design. Whereas the fixed barn machinery of the north was necessarily the property of farmers, the small portable machines were often owned by a new group of users, threshing contractors. Machines of this era were literally just threshers. The major difference between using the two was that the portable machines were not linked to the separate "rakes and fanners" of their larger fixed brethren.

Fig 281: Lester's "London" Thresher of 1802

In 1839, Richard Garrett exhibited a horse-driven thresher at the South Devon Show, Exeter. It was claimed that the machine could thresh between 180 and 225 hundredweight in a day and Garrett was awarded a prize. Unfortunately, no records exist as to the crop being threshed or the number of horses required to drive the machine.

Fig 282: Portable Thresher in Transport position (Garrett)

In 1841 at the Cambridge Royal Show, a total of 8 portable threshers were exhibited. The two machines built by Ransome & May, and Garretts were the only ones to successfully complete their trials. They were said to have been capable of threshing "25 to 30 bolls" in a day, equivalent to 150 to 180 bushels. The winning Ransomes machine weighed 35 cwt and featured an unconventional disc-type rotary steam engine to a patent of Henry Davies of Birmingham powered by a cylindrical boiler with a hemispherical top mounted vertically. Overall design of the machine was by Ransomes' Works Manager, William Worby. The boiler and motion were carried on a 4-wheeled wooden wagon-type chassis extended to provide a platform onto which the small thresher could be loaded for transport. This machine is frequently referred to as being the very first self-moving portable. A well-known contemporary drawing shows the chain drive to a single rear wheel.

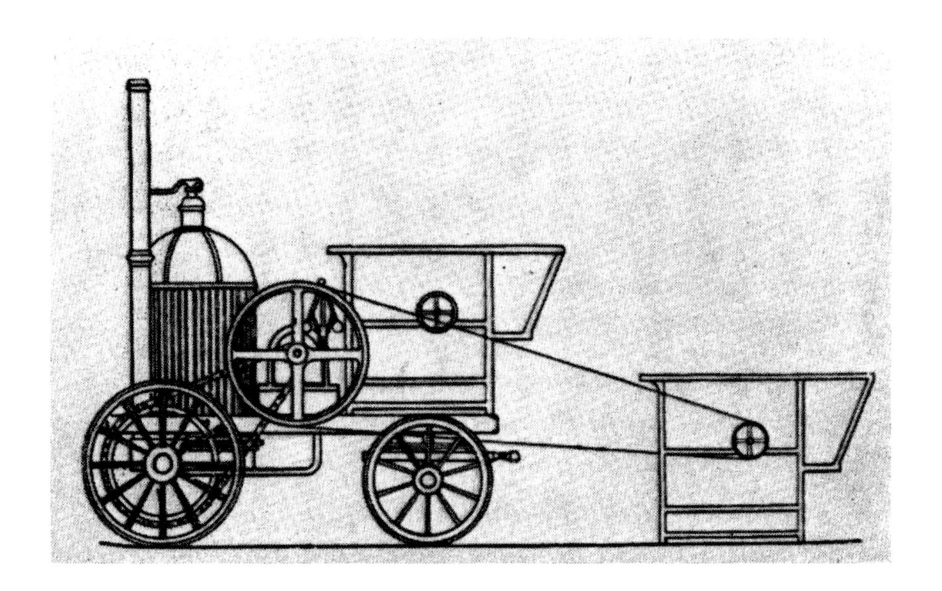
Fig 283: Ransomes Self-propelled Thresher

What is less well-known is that the machine, reported to be capable of 4 mph, ran away at the 1842 Bristol RASE Show and broke a fence. One wonders if the horse used for steering was in the shafts at the time. The accident didn't prevent the awarding of the top prize of £30 in the miscellaneous class. This drawing also shows that while actually threshing, the small thresher stood on the ground and was driven by a belt. A press report of the time said that "it was now possible for the whole threshing process to be carried out in the field next to the stacks". The thresher itself was based on a design that originated in 1802. The body of the drum was a wooden barrel 3' 6" long and of 19½" diameter. The surface was completely sheathed in cast-iron plates with ribs projecting ¼" in a criss-cross pattern The wooden concave was mounted on top of the drum and covered about a quarter of its circumference. The concave was mounted on short chains which allowed it to move thus avoiding blockages. Drums of threshers of the time commonly carried 4, 5, or 6 beater bars.

Even at this late date, not all development work was centred on mechanically-powered threshers. A Mr Joseph Barling of Maidstone, Kent received a £5 prize at the 1842 RASE Bristol Show for an improved design of hand thresher which was built for him by Ransomes. The majority of similar machines were constructed so that pairs of men could turn two cranks at either end of a shaft, one pair on each side of the machine. The drum was gear driven off the shaft on a ratio of around 8 to 1. Mr Barling's design substituted a bar and ratchet arrangement for one of the handles so that the pairs of men could change over from time to time. But one wonders if this arrangement really does prove that a change is as good as a rest.

A more significant design change was demonstrated by Henry Crosskill of Beverley. His thresher featured a novel arrangement whereby the clearance between drum and concave could be adjusted while the thresher was still operating. Before this, adjustment was usually achieved by 2 pairs of screws, although Messrs Barrett, Exall, & Andrews were offering a concave on an eccentric mount that used "worm-gear working on a toothed circumference". This arrangement was considered better when larger clearances were required, such as when threshing peas and beans. A notional starting clearance for cereals on these early machines would have been around 1½" at the feed point narrowing to 1" at the exit from the drum. An article dating from the early 1850's warns that "Bad threshing is generally the result of too much meddling with the drum clearance adjustment by the ordinary farm labourer". Exactly the same sentiment is echoed many times over the next 100 years.

In 1842, William Howden exhibited a threshing machine at the Lincolnshire Agricultural Show held at Wrangle north of Boston. This was seen as a major grain-producing area; in 1811, it was stated that one third of all grain for London was shipped from Boston. Howden's thresher was described as "a simple machine without shakers or dressing" and "so arranged to be powered by steam or horse-gin". Howden had already built a number of portable engines to power screw-operated drainage pumps and a local farmer named Smith persuaded Howden to use one of these portables to drive the thresher, not by the customary belt, but via what was described as a "coupling shaft".

Fig 284: Tuxford's Self-propelled Thresher

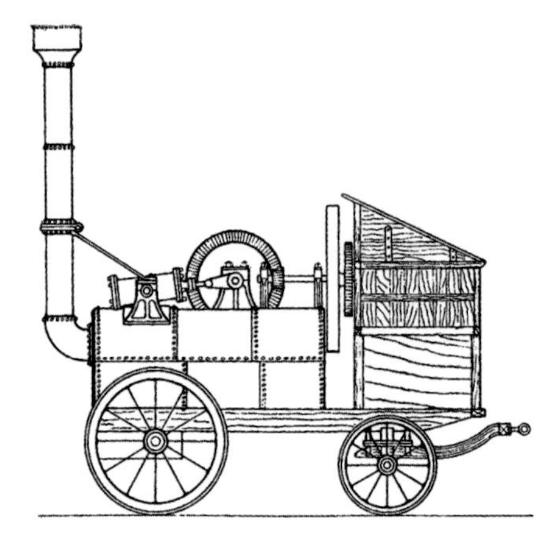

Also in 1842, William Tuxford produced a combined portable and "finishing thresher" to a design originally suggested in 1839 by a Mr Wingate of Hareby, and John Morton, the manager of Earl Ducie's Whitfield Experimental farm in Gloucestershire. Like the earlier Ransomes machine, power source and thresher were both mounted on a wooden 4-wheeled chassis. Coincidentally, the Tuxford also featured an unconventional steam engine. This time this was in the form of an oscillating cylinder rather in the style of a Mamod model and known at the time rather unflatteringly as "the monkey on the bear's back". Unlike the Ransomes, the machine was designed to operate as a single unit when threshing, main drive being taken through 90 degrees via a large bevel gear. As with most of its contemporaries, transport from place to place was to be courtesy of a team of horses. The first of an initial 7 sets was purchased by one Robert Roslin of Algarkirk, the machines being rated at 6 nhp and weighing 3 ¼ tons. The combined single unit was not seen as wholly successful. The remaining machines of the first batch plus a second batch of 12 machines whilst retaining most of the features of the original design kept thresher and engine separate. The specification of the thresher part of the set was very advanced for its time and could be described as a "finishing" machine featuring as it did, a drum and concave, a shoe with a fan, and straw shakers. Unusually the latter were in the form of a revolving drum rather than of reciprocating pattern in the manner of a miniature Scotch machine. This was in contrast with most machines of the same date, such as the Ransomes, where the only moving part was the drum, a rigid frame holding a feed board, the concave, and a second slatted board for the straw pass over.

The small firm of Hensman went against the popular English trend of the time by offering a type of peg-drum thresher. Their machine was so designed that although both drum and concave were fitted with pegs of "Tooth" or "Vandyke" pattern, the drum clearances were so great that the pegs did not interlock, thus reducing the damage to the straw, and at the same time, making a machine where the crop offered less resistance and hence required less power to drive.

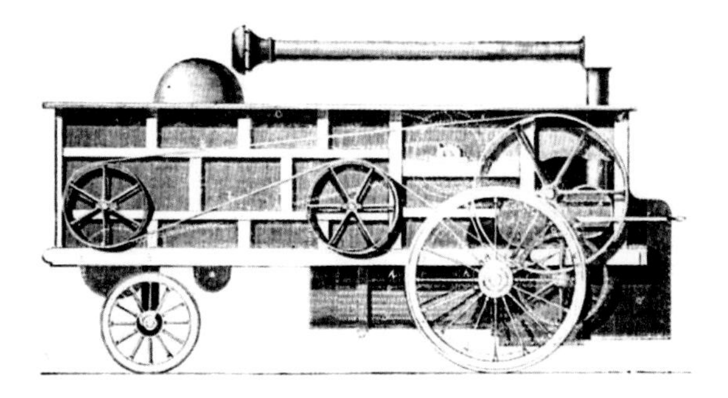

Fig 285: Crowek's Self-propelled Thresher

In spite of the fact that incorporating the thresher and its motive power had not found favour, Daniel Crowe of Gaywood near Kings Lynn produced a few machines to this design in his Victoria Steam Works as late as 1867. A return-tube boiler supplied a steeple-type engine which by another coincidence was the same arrangement as used by later Tuxford engines. The thresher part was completely conventional and at the Leicester Agricultural Show of 1868, the 7 ton 1 cwt machine successfully threshed 5½ cwt of what was described as "inferior wheat" in 9 minutes.

Returning to 1843, Lincolnshire Contractors had already set a tariff for threshing work. The RASE Journal quoted figures of between 1/ and 1/3 per quarter of grain threshed, but these excluded the cost of labour and coal.

The 1846 Ransomes catalogue still showed a simple thresher but this could be used in conjunction with their separate winnowing machine. The latter featured reciprocating sieves driven from a crank on a single shaft which also carried a fan which directed a blast over the sieves. The thresher used a simple fixed feeding board set an angle of 30 degrees towards a point midway between the circumference and the centre of the drum, which was built as a "skeleton with beater bars". The concave covered a third of the circumference of the drum and was made up from iron webs and "open wirework". The drum clearance could be adjusted by screws. Typical settings would have been 1½" at the feed point at the top tapering to 1" or even ¾" at the bottom. This would be very similar to the settings on a modern machine. The threshed material fell onto a fixed "harp" or riddle where grain and stalks would be separated manually using rakes and forks. The size of the thresher was around 6' 0" by 4' 6" and was designed to be transported on a purpose-built 2 wheel trailer which also carried the associated horse-whim. Criticisms levelled at this type of machine were that they were known to break straw and to bruise or "nib" barley. James Allen Ransome was staunch in his support for the machines produced by his company stating that many of the problems stemmed from operation by "less-skilled operatives" and that many machines of this type had been assembled from individual components by "less-able firms". The bruising of the straw was defended by Sir John Sinclair as making it immediately more suitable for use as fodder. Mr Ransome was quick to point out that this would be better left to the use of a chaff cutter.

Another major step forward.....

Fig 286: Garrett "Improved" Thresher

In 1846, Garrett exhibited their "Bolting" thresher at Shrewsbury Royal Show. This featured what they described as "Parallel Movement" which allowed the feeding of sheaves or "bolts" crosswise rather than the more normal endwise. In order to achieve this, the width of the machine had to be substantially increased. This system produced, for the first time, unbroken straw that was suitable for thatching. Another advantage quoted was that the quality of the straw made it more suitable for resale, particularly for use in urban areas. A contemporary report said that the straw could be "more easily handled, particularly after introduction of a straw tier". Their catalogue guaranteed this 4 hp machine would thresh over a ton of grain per hour. The earliest of these machines featured a drum built as a solid cylinder clad with iron plates fitted with 8 bars of ¼" projection and a concave that covered 60% of the circumference of the drum. The clearance between drum and concave was less than on most machines of the period so, to prevent clogging, an automatic feed apparatus was fitted consisting of a pair of fluted rollers. In view of the damage caused to straw by these rollers in the past, one imagines that those feeding this machine were instructed to take particular care. The original machine could also be adjusted to thresh beans by reducing the rotational speed and increasing concave clearance to accommodate the fitting of just 2 bars each of 1" projection.

The later machine exhibited at the Show was of a more modern layout featuring Garrett's patent cylindrical drum and wire concave the latter derived from the original Ball design. Although Garretts are generally credited as being the first to offer the "bolting", thresher, Ransome and May are quoted as having exhibited a machine which offered crossways feeding at the 1843 Derby RASE Show. By the time that Fosters produced their first machine in 1848, there were at least 25 manufacturers of threshers.

Yet another major step forward...

In 1848, Burrells won a silver medal at the Royal Agricultural Society of England's York show with their latest design of threshing machine. This was officially known as an "Improved Patent Threshing, Shaking, Riddling, and Winnowing Machine" and may be said to be the first modern threshing machine as they are thought of today. An improved model won a £5 cash prize at the Bath and West of England Society's Show of 1853, and the judges at the RASE Chelmsford Show were unanimous in their praise. They stated that it had done "perfect work" with "no corn broken, or where it ought not to be" and achieved "perfect separation of chaff from cavings". As well as its operation, the appearance of the machine is a clear precursor of the mainstream of thresher design albeit on a slightly smaller scale. The crop was fed in at the top direct to the drum and wire concave and the straw was taken to the end of the machine by a set of walkers. The grain fell into a set of screens in an oscillating shoe with an air blast from a full-width fan mounted at the bottom of the machine. The rough-dressed grain could be drawn off directly at this point or allowed to fall into a hopper at the base of an elevator to take it to the top of the machine and the "Second Dressing Apparatus". Unlike a modern machine, this was turned through 90 degrees but otherwise the internal layout was basically to remain the same for the next 100 years. The grain from the elevator passed through the awner and fell into the riddlebox where it received an air blast from a smaller, separate second fan. It then fell into a grading screen where it is diverted to 3 corn spouts, one for tail corn and two for best. The price of this machine was £115, with the optional reciprocating screen costing an extra £5. This was also a Burrell design covered by patent no 2757 of 1804. It is interesting to note that a number of other manufacturers, Humphries and Garretts to name but two, produced machines with this 90 degree change of direction. It seems an unnecessary complication when compared with the all-parallel shaft arrangement used in modern machines, and yet this layout was revived for "bagger" combine harvesters of the 1950's.

Fig 287: Burrell's first "Combined Thresher"

In 1848, John Goucher of York took out a patent for drop-forged beater bars for the drum. A later patent taken out in 1854 was for changing the profile of the bars to diagonally-fluted. The flutes provided what were described as "clearing spaces" for the grain resulting in far less damage. Some manufacturers fitted these bars with the fluting alternately left-to-right and right-to-left thus achieving greater agitation of the crop. The clearing spaces were provided on the bars of some early machines by the simple expedient of winding a coil of wire around a rolled steel core.

Great emphasis was placed on the methods used to secure the ends of the wire to prevent it from coming adrift when in operation. In spite of these improvements, most English and Scottish threshers were fitted with flat bars well into the 1860's. In 1849, a Garrett machine was described as being the "First portable thresher to give a more or less clean sample of grain". A Mr W Robinson of Belfast designed a machine, based on then current thresher practice, for the separation of seed. It was exhibited at the 1851 Smithfield Show by Deane, Dray & Co. and was claimed to separate rye-grass from foil grass, black seed, and other small seeds. It used a reciprocating angled screen together with a series of wire gratings of various sizes, and was also claimed to be suitable for flax seed and corn.

Across the Atlantic in America, the brothers Hiram & John Pitts followed their patent for a horse treadmill in 1830 with another for a "thresher-cleaner" in 1834. This was followed in 1837 by their famous "apron conveyor" thresher where the reciprocating walkers were replaced with a board raked by horizontal bars on a pair of endless chains on sprockets. This model is often quoted as being the first American "combination thresher" with integral winnowing. The later threshers produced at a new factory in Chicago from 1852 were christened "Chicago Pitts" and these, worked by a team of 8 horses, were capable of threshing between 300 and 500 bushels of wheat per day. This factory was run by Hiram Pitts; the brothers' partnership had split up by this time; and John had joined Joseph Hall to produce the "Buffalo-Pitts" threshers in Rochester, New York.

In 1850 Burrell produced a revised design of thresher which featured a single blast. This design continued up to at least 1875 when a 4' 6" machine was priced at £135 and the larger 5' 0" model at £145. This was considerably cheaper than the corresponding double blast machines that cost £160 and £170 respectively. Examples were exhibited at the RASE Show of that year and competed against machines built by Garretts and Barrett, Exall, & Andrews.

As late as 1852, G H Andrews in his "Rudimentary Treatise on Agricultural Engineering" stated, "Eminent agriculturalists are divided in opinion as to whether it is better to have a fixed engine and the barn machinery fitted up in a building and bring the corn to the machine, or have a portable engine and threshing machine and take the machine to the crop." The author seems in no doubt seeing the fixed equipment as being "cheaper to run and requiring less repair." He also offered the following advice to manufacturers: "Machines to be driven at high velocities should be better framed." Sir William Tritton, then Managing Director of Fosters wrote in 1935, "Essentially the system of the 'combined thrasher' is the same today as it was in 1854". Despite this, there were always a number of alternative detail designs being tried such as Richard Hornsby's auger-type grain pan mounted below the concave and walkers of one of his machines built in 1856.

As an aside, in 1857 Charles Burrell filed patent number 917 dated 12 April for an "Apparatus for screening Corn & Seeds". This consisted of a screen suspended at an angle by a set of 4 swinging links actuated by a crankshaft and connecting rods. Below the screen were arranged a series of rods arranged in pairs with scraper blades separated by washers so as to match all the gaps between transverse wires of the screen above. A revised Charles Burrell thresher design was the subject of patent 1344 of 5 November 1867 entitled "Improvements on Drawing and Dressing Seeds". The design included a riddle to remove stones, grass, and other seeds, a sheller, comprising a revolving drum carrying inclined beaters within a conical barrel with a grooved interior where the shells are rubbed off the seeds, an elevator, a second or dressing riddle with an air blast at the end to remove chaff, and a second sieve arranged to direct shelled seed to one spout, and unshelled to another.

The 1858 RASE show was held at Chester and, for the first time, full thresher trials were held. No fewer than 55 machines were demonstrated. But the judges were not impressed. They noted that the majority of machines "could not complete their work without damage to the grain, or corn left in the cavings and chaff." and of the 55 machines tested, they felt that they could give their approval to only 4.

In 1859, Garretts produced the first of their popular "single fan" threshers, one of which was later demonstrated at the 1862 Great Exhibition. This design was the subject of patent number 153 in the joint names of Richard Garrett and James Kerridge, then the foreman of their threshing department. The machine was arranged to carry the single large-diameter narrow-width fan keyed onto the end of the drum shaft thus reducing the number of high speed rotating parts to one. All blasts, at first two, and later three, were directed by ducting and this economical design stayed in mainstream production until 1875.

Fig 288: Garrett Single Fan Thresher as exhibited at the exhibition 1852 - Elevation & Section

At the 1862 Battersea show, one of the judges Harry Evershead, made an interesting comment regarding the life expectancy of threshing machines. In his opinion if a double blast machine could last 8 years, then a single blast machine ought to last 10. A mark of their popularity was that production of the single fan design did not finally cease until after 1900, although by that time, the multi-fan design had become very much the norm. The simplicity of these machines must have been offset by the poor efficiency of the rectangular-section wooden ducting which would have resulted in considerable transmission losses. These could have been greatly reduced had later models been fitted with cylindrical ducting in light-gauge steel but although improvements were made, the single fan system was eventually seen as inherently less efficient than the multiple individual fans with their short ducts.

The first designs for straw shakers used a pair of crankshafts. A number of manufacturers, notably Clayton & Shuttleworth, produced machines with a single central crankshaft. These machines usually carried 5 shaker frames instead of the 4 used on twin-crank machines and these were hinged on short rocker arms, alternately at the drum and straw outlet ends. Although these single-crank machines enjoyed considerable popularity, eventually ALL manufacturers turned to the double crank system.

In search of an even greater reduction in the number of moving parts, the Wantage Engineering Company produced a small machine named the Simplissima. The shakers were driven off a single central multi-throw crank which also operated both shoes. A further reduction in moving parts was made by mounting the dressing fan directly on the awner shaft.

Some manufacturers including Nalder & Nalder and Brown & May elected to mount the riddle crank below the shakers. This made for longer connecting rods but had the advantage that the angle that the rod was put through was less. This reduced the stress on the rod itself which had to flex because of the rigid connection to the shoes at the "small end".

The RASE Thresher trials held were generally very well supported. A notable exception to this was the shows at Canterbury in 1860 and Leeds in 1861 which were boycotted by many of the manufacturers following a dispute with the organisers.

Chapter Fifteen

MOTIVE POWER 1 - BEFORE STEAM

Thresher Nameplate (C & S)

The modern threshing machine is a compact, if somewhat large wooden box mounted on wheels for easy transport. But the earliest machines were huge pieces of fixed machinery built into a barn.

Water Wheels and Windmills

Fig 281: Lester's "London" Thresher of 1802

The earliest record of a water wheel is Greek and dates from around 500 BC. Many of the earliest threshing machines relied on water wheels for power just like flour mills on whose form of construction they were based. Originally these would have been built entirely from wood but this later changed to cast-iron with wooden fittings. Choice of woods played an important part in their design. For example, if Lignum Vitae was used for the bearings because of its hardness, then apple, beech, or hornbeam were used for the teeth because these woods are all naturally slippery. The wheels could be arranged in two major ways, overshot or undershot. An alternative was the water turbine but this was a much later development. Although more expensive in terms of construction costs, the overshot wheel was seen a the more efficient design being able to translate 75% of the available energy into useful power.

Fig 290: Water Wheel - Overshot pattern

Unlike water wheels, windmills were not generally seen as a viable power source for threshing. Despite their first recorded use in Britain in 1191, an article written in 1803 described these as "an uncertain power" and an article of 1852 is even more scathing. Here the author describes them as "exceedingly expensive both to construct and. maintain, uncertain and irregular in their action". Interestingly, Andrew Meikle took out a patent for adjustable shutters for windmill sails with a view to solving this very problem.

Horse Power

Mention has already been made of animals being used for threshing, first by simply walking over the crop and secondly by dragging frames or rollers round and round. A natural development of this was the "Horse Sweep". Here the horse was harnessed to a beam that pivoted around a fixed post. Similar machines were illustrated and described by Agricola in Germany in 1556 and by Ramelli in Italy in 1588.

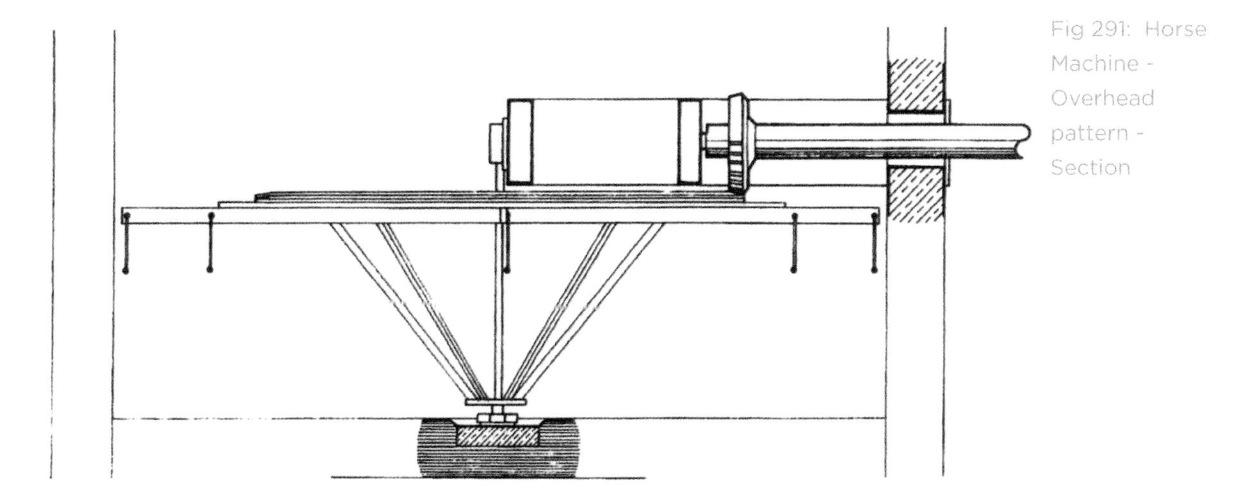

Fig 291: Horse Machine - Overhead pattern - Section

This system developed into capstans known as a horse-wheel. A large concentrate gear-wheel attached to the vertical slow-rotating beam meshed with a smaller cog and this drove a smaller horizontal shaft that could be connected to anything requiring rotary motion. These capstans fell into two distinct types, overhead or underfoot. Overhead machines were of a permanent fixed design often within a purpose-built enclosure or round house.

Fig 292: Horse Machine - Overhead pattern - Roundhouse

As its name suggests, the overtype's concentrate gear driving the small horizontal shaft was at high level, with the horses, sometimes as many as 4, turning the beams of the heavy capstan suspended from yokes suspended from above. One particular design by Walter Samuel, a blacksmith of Niddrie, just east of Edinburgh, used a system of rope loops through pulley blocks to ensure that the load was evenly distributed to each horse. His whim was based on the horses walking an 18' diameter circle to drive a pair of bevel gears giving a ratio of 4.9: 1.

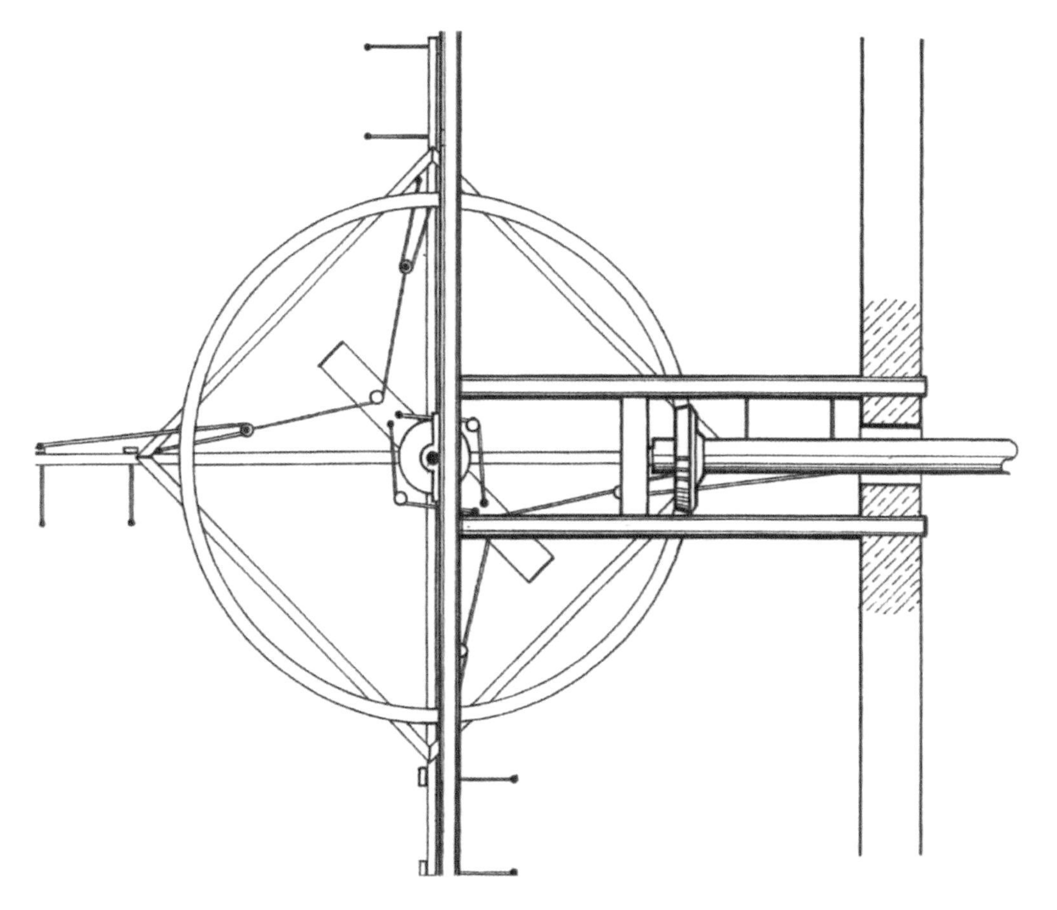

Fig 293: Horse Machine - Overhead pattern - Capstan (Niddrie)

A similar arrangement described in "Transactions of the Highland Agricultural Society" designed by a Mr Christie of Fifeshire, sought to prevent one or more horses "hanging back" by the use of his "ring-chain" arrangement. Mr Christie also advocated the use of a "swingle tree" rather than a simple yoke because otherwise, the horse would be "exerting itself tangentially".

By comparison, underfoot machines were usually smaller and often used out of doors. In theory they were mobile except that the base for the centre post had to be somehow firmly fixed to the ground. One, or at most, two horses pulled a single beam via trace chains rotating the centre post with similar gearing to the overhead machine, but this time, at ground level. Where the horses walked over the small horizontal drive shaft or "tumbling rod", a cover was often provided. The step-up gearing required was provided by a large concentrate wheel driving a small cog wheel. Operating close to the ground, it was not uncommon for a tooth on the concentrate wheel to be broken off when a stone kicked up by the horse became jammed in the mechanism. Garretts received an RASE prize for their improved design of horse machine where the concentrate wheel was constructed in a number of segments each of which could easily be replaced in the event of a damage.

Fig 294: Horse Machine - Underfoot pattern

An 1852 article sees "Horse-gear" as being only suitable for use on "exceedingly small farms" and that "Now it much more economical to use steam power if 4 or 5 hp is required". By this date, the smaller and cheaper underfoot machines were far more popular option than the overhead system. At the 1867 Bury St Edmunds Royal show, the judges proposed that no further awards should be given to these machines saying that they were "old-fashioned" and "represented a sign of backward agriculture". But although mechanical power in the form of steam, was in the ascendancy, Garretts exhibited a horse-powered thresher at the 1874 Smithfield Show, and as late as 1880, a patent single horse "Horse-power machine" built by Crowley received a silver medal at Carlisle Royal Show This was designed to provide an output of 4 nhp running at 100 to 120 rpm via a gear ratio of 37:1 meaning that the lap time for the horse was thought to be around 20 seconds. This late date demonstrates that barn threshing was still very popular in Northern England and Lowland Scotland.

Fig 295: Horse Machine - Underfoot pattern - 1880 (Crowley)

Fig 296: Horse Machine - American Treadmill

One alternative design to the horse-wheel was the horse treadmill. This was fully mobile being mounted on wheels for easy transport but suffered from the disadvantage of very many more moving parts. To some extent, the amount of power developed by the treadmill could be controlled by adjusting the angle of incline of the track. However, too steep an angle would put strain on the legs of the horses. This problem was solved by the development in America of the "Level Tread" design, rather akin to a modern escalator. Unfortunately, this suffered even more from the original problem of too many moving parts. Nonetheless, a generation of early threshers in North America, were driven by these machines. However they were virtually unknown in Britain.

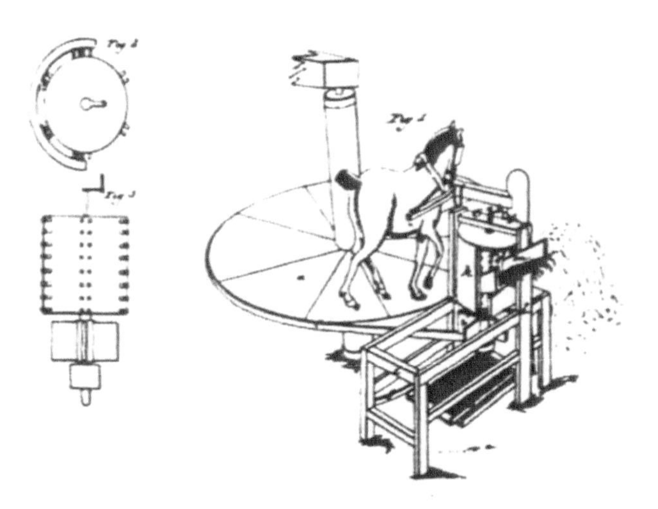

Fig 297: Horse Machine - American Roundabout

Fig 298: Full Circle - Roundhouse converted as holiday accommodation

Table 8: Contracting Prices

PRICES FOR 1950 / 1951 SEASON				
Per Hour	**Men**	**£**	**s**	**d**
Threshing: Traction Engine + Machine + Chaff Cutter	2	1	2	8
Threshing: Traction Engine + Machine + Elevator	2		18	0
Hulling: Traction Engine + Machine	2	1	0	0
Bailing: Traction Engine + Machine: Wire	2	1	8	4
Bailing: Traction Engine + Machine: Wire	1		7	0
Chaff Cutting: Traction Engine + Machine	2		7	0
Extra: Trusser			2	6
Extra: Self-Feeder			1	6
Per Quarter		**£**	**s**	**d**
Threshing: Hard Corn			9	0
Threshing: Oats			8	6
Extra: Bailing			2	6
Extra: Chaff Cutting			1	0
Extra: Straw Tying				10

Chapter Sixteen

MOTIVE POWER 2 - STEAM

McLaren Trademark

The Engineman's eleventh Commandment:
"Six days thou shalt labour, and on the seventh day, thou shalt wash out the boiler".

By far the biggest change in threshing came with the introduction of steam power. For the purposes of this book, the steam engine as used to drive the threshing machine may be seen as belonging to one of four groups: stationary engines, portable engines, traction engines, and steam tractors.

Comparatively few stationary engines were used to drive threshing machines. The first mentions of steam-powered threshers relate to a machine used at Trentham, Staffordshire, in 1796 and another by one John Wilkinson in 1798 in Denbeighshire, Wales. A machine is reported as being in use in East Lothian in 1799 and yet another by a Col. Buller in Heydon in 1803. This engine was arranged to also power a chaff cutter and a grinding mill, simultaneously when so required. The cost of this installation was believed to be around £600, an enormous sum when a year's wages for a ploughman would have been a mere £12. Another installation at Chillingham Barns in Northumberland of a slightly later date cost £100 for the thresher and £325 for the engine although this included the building that enclosed it. The thresher featured a 6' wide drum and was claimed to be able to process 1,000 to 1,200 sheaves an hour.

An article written in 1852 states that the first condensing engine in Scotland was built in 1825 and was still working. Similar machines were still being built to the same design.

In a paper given to the Institution of Mechanical Engineers in 1856, the author, William Waller listed the advantages of the portable engine over a fixed installation particularly with regard to threshing. He mentions savings in the cost of barns and reduced transport costs if the threshing can be brought to the crop. The fact that many of these farms were tenanted, meant that the machinery was owned by the tenant rather than staying with farm as the property of the landlord.

Notwithstanding the use of the horse power machines and the stationary engines described above, it could be said that it was the portable that made the operation of a modern threshing machine possible. Indeed their two histories are inextricably interlinked. Threshing by steam has been described as "the first application of mechanical power in agriculture".

High-pressure steam as opposed to a vacuum generated by condensation was first proposed by Leopold in Plunitz, Saxony in 1727 for use as a pumping engine. It was not intended to provide rotary motion and it is doubtful if his design was ever built. In 1780, John Pickard took out the patent for the simple crank. This forced James Watt to devise his much more complicated Sun-and-Planet gear of 1781. Watt's pupil, William Murdoch took out a patent for a slide valve in 1785, and Edward Cartwright greatly improved efficiency when he patented metallic packing and piston rings in 1797. As an aside, the credit for the design of the earliest automatic valve gear used on these Watt engines is given to a boy named Humphrey Potter. It was named a "scogger" gear because the boy, instead of operating the valve manually, could "scog" or skive off.

Portable and thresher were moved from place to place with teams of horses. Steam transport had already appeared, not only on the railways but also on the roads. There was a brief period in the 1820's when a number of steam road coaches were run on the roads before they were killed off by restrictive legislation and punitive tolls. These could be up to 12 times those levied on horse-drawn vehicles. It seems strange now to think of steam engines being towed from place to place by teams of horses. This is partly because it is just as strange to visualise a world with millions of horses serving all local transport needs.

Fig 299: Trevithick semi-portable engine of 1811

As a Cornishman, the author is particularly proud to be able to say that these steam road coaches were predated by Richard Trevithick's steam carriage of 1801. Cornish people still sing "Going up Camborne Hill coming down"; the statue of the great man standing in front of the library surveying his original route. Cap'n Dick, as he is affectionately remembered invented the blast pipe, the fusible plug, and introduced heating for the cylinder. More importantly in the context of this book, he is also credited with designing the first practical portable engine in 1812. It was described as a high-pressure portable engine for agricultural purposes mounted on wheels. It weighed 15 cwt and sold for 60 guineas. An earlier engine built by Trevithick in 1811 for Sir Christopher Hawkins of Home Farm, Trewithen, Probus, Cornwall which cost 80 gns was exhibited at 1879 Kilburn Royal Agricultural Show's "Museum of Antiquities". The machine had been working for over 60 years and one of its duties is recorded as that of threshing. It was taken apart for transport, re-erected at the showground, demonstrated at the show, taken to pieces again then taken to the Science Museum at South Kensington where it still resides. Perhaps this design of engine would be better described as semi-portable but they were used in Cornish mines as drive for "whims" and the Cornish miners abroad would order them to be sent out in pieces for re-erection. Interestingly, some more modern portables were also designed to be demountable rather than being mounted on wheels.

These early Trevithick engines operated at over 30 psi, a far higher pressure than that used on Watt's machines. Boulton and Watt even attempted to get an Act of Parliament passed banning such machines on the grounds that they posed a safety risk to the public. Trevithick uprated a number of existing Watt engines including one at Wheal Prosper mine, Gwithian, less than 5 miles from the author's home. Here the boiler pressure was raised from the original 5 to 10 psi to an intimidating 40 psi! Unfortunately this proved a step too far and the boiler exploded fortunately without causing any injuries. Despite the concerns expressed by Messrs Boulton & Watt, the longevity of early Trevithick-designed machines is remarkable and the Metropolitan Water Board's 90" engine built at Copperhouse, Hayle in 1846 worked continuously at Kew Pumping Station until 1943, where it remains in preservation.

Fig 300: Late-pattern Stationery Engine

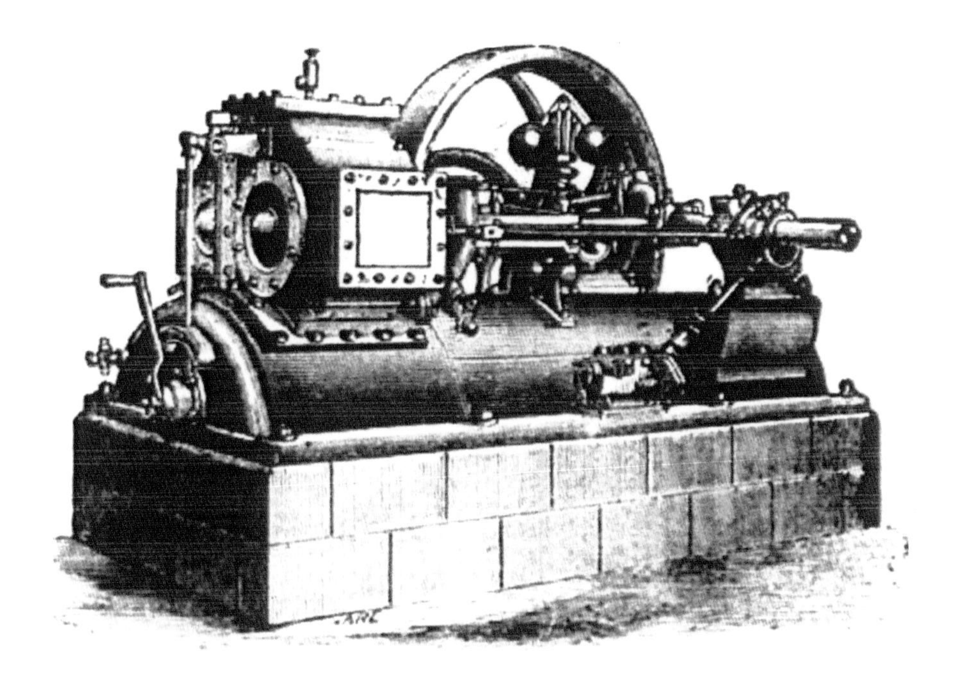

In an article written in 1912, mention is made of some condensing engines being used in Scotland to power fixed equipment. It goes on to say that "Fixed thrashing machines are still employed to a much larger extent than in England".

In 1802, one Matthew Murray had filed a patent for an engine "transferable without being taken to pieces" which could be used for "any process or manufacture requiring circular motion, or for the irrigation of land, or for the various purposes of agriculture" but it seems unlikely that the good Mr Murray's ideas were actually transferred into wood and metal. Mention should also be made of the other pioneers of the era, Bramah, Hornblower, and Wansbrough. But little practical progress was made over the next 30 years due to the unrest in the agricultural population following the Napoleonic War. There are records of a portable design by Nathaniel Gough of Salford dating from 1830 and from the same era, designs by several others including Alexander Dean of Birmingham, and Colbourne Cambridge of Market Lavington, Devizes, better known as the originator of the Cambridge Rolls.

Fig 301: Portable Engine - Layout

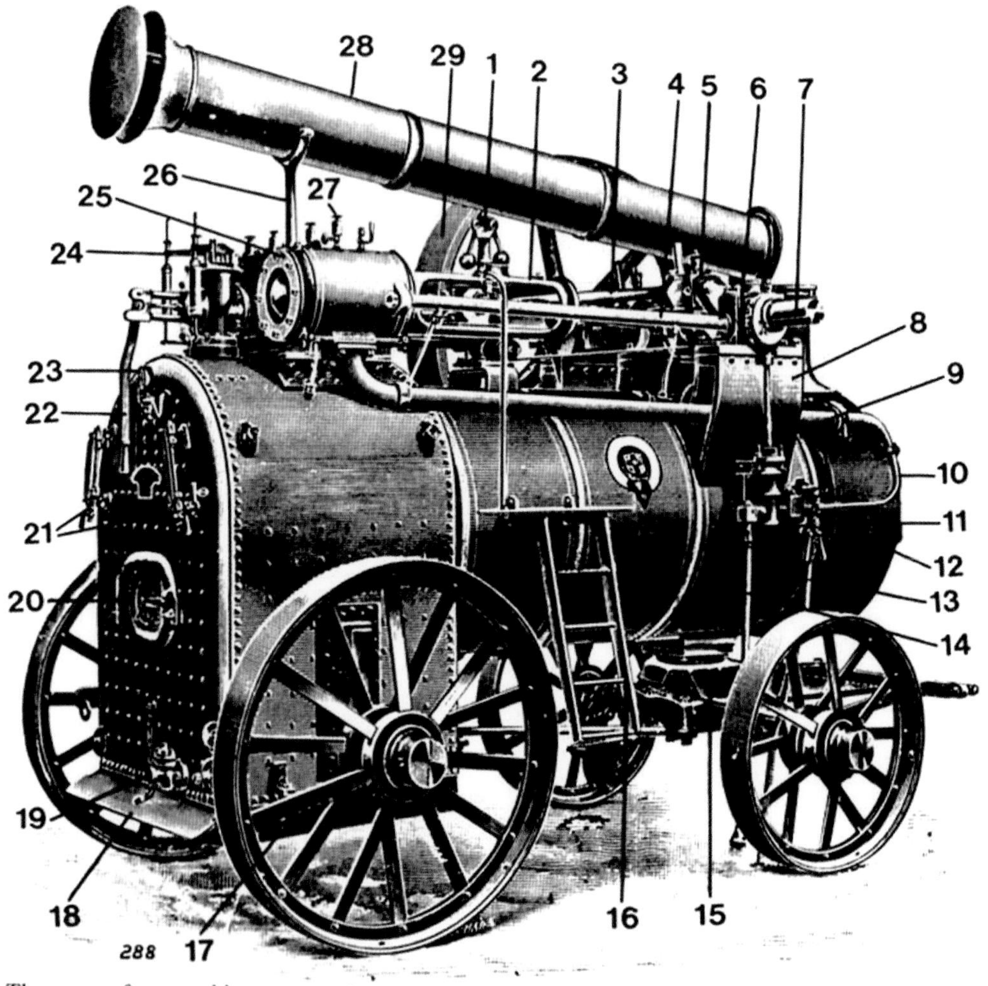

The parts of a portable steam engine.

1 Governor
2 Crosshead guide
3 Connecting rod
4 Tie rod between cylinder block and crankshaft bearing
5 Big end
6 Crankshaft bearing
7 End of crankshaft
8 Crankshaft bearing bracket
9 Exhaust steam pipe
10 Branch pipe taking some exhaust steam for feedwater heating
11 Smokebox
12 Boiler feed pump
13 Return pipe from boiler feed pump
14 Suction pipe to boiler feed pump
15 Swivelling forecarriage
16 Access ladder for oiling and maintenance
17 Ashpan
18 Damper allowing air to enter under fire
19 Drain cock for draining boiler in frosty weather
20 Firehole door
21 Water gauges
22 Regulator lever
23 Steam pressure gauge
24 Two spring balance safety valves
25 Cylinder block
26 Rest for chimney
27 Cylinder oil cup
28 Chimney, folded down
29 Flywheel rim

The major manufacturer, Clayton & Shuttleworth, produced their first 2 portables in 1845, and their 1846 design was the first of what became the conventional layout with motion mounted on top of a locomotive boiler. At the1847 Northampton Show, 7 portables in a row competed head-to-head driving threshers to thresh a given number of sheaves against the clock. As the amount of material available was small, the best time was a mere 8 minutes. The Royal Agricultural Society of England did much to promote the use of portable engines with the trials held at their annual shows. However concerns were expressed that these were "racing machines" built especially for competition with boilers having very narrow water spaces and a large number of thin-walled tubes, and requiring expert driving and maintenance. They also tended to use best Welsh coal at a time when Yorkshire coal would have given a more accurate indication of what could be expected in normal use. Requests were made to increase the length of trial and try to have the manufacturers compete with production machines. The RASE had fixed a maximum safe pressure for its earliest shows of 45 psi but by the mid 1850's, pressures of 150 psi or more were not uncommon.

Fig 302: Portable Engine in transport position

Fig 303: Typical Portable Engine

By 1851 it is estimated that over 8000 portables had been sold. As the agricultural world gradually became more settled, thoughts began to turn towards harnessing the power of the portable for locomotion. This was first achieved by the simple expedient of driving one rear wheel directly from the crankshaft by a length of the recently-developed pitch-chain. Thomas Aveling modified a series of 20 portables which were built for him by Clayton & Shuttleworth and fitted with a tensioning device for the chain which was the subject of Aveling's patent no 1995 of 1859. A tender was attached to the rear of the engine, firstly running on its own wheels as in railway-engine practice, and later by constructing a platform extending from the rear of the firebox of the engine itself. The slide valve to control admission and exhaust of the steam to a double-acting cylinder was operated by a simple eccentric on the crankshaft. The engine could be made to run in reverse, but this entailed the use of spanners in order to re-set the die-block. An easier option when driving machinery was simply to use a crossed belt.

To allow manoeuvring, this simple valve-operating system was replaced by designs borrowed from railway practice, by far the most widely used being Stephenson's Link Motion designed by William Williams and William Howe in 1842. The valve itself was generally the same slide type as used on the portable but some manufacturers used the piston type valve more common on railway locomotives.

Steering was by a horse in shafts. This seems curious today but it should be borne in mind that a horse would still be required to carry out a number of tasks associated with threshing, not least of which would be pulling the water cart. A second reason was that it was thought that a horse preceding one of these machines could lessen the risk of other oncoming horses shying. This level of concern also resulted in legislation that prevented the use of any steam engine within 25 yards of the highway unless it was hidden from view. This restriction lasted until the 1894 Locomotive Threshing Machines Act. Eventually the horse in shafts was replaced, first with a system patented by Thomas Aveling known as Pilot Steerage, patent no 891 of 1860, where a man sat in front of the smokebox operating a tiller arm connected to a narrow 5th wheel pivoting in a frame where the horse had previously stood. This in turn gave way to a steerage platform in front of the smokebox where the steersman turned the front axle with a ships wheel. The final change was to move the steering back to the tender by the driver. But the principle of separate driver and steersman was to remain. An exception to the above was the "Farmers' Engine" built by T B Willis to the design of Wiliam Worby and exhibited by Ransomes in 1849.

The portable's 5' 0" to 6' 0" diameter spoked flywheel, incidentally weighing over 6 cwt, had originally been mounted over the smokebox, next to the machine it was to drive, with the cylinders and motion over the firebox. This layout had the additional safety advantage that the motion was kept away from the operator. Governed speed was from 100 to 120 rpm with a maximum pressure of 120 psi. The self-moving portable made the final metamorphosis into the traction engine when this layout was reversed bringing the crankshaft nearer to the rear axle thereby allowing all the drive to be by gearing and doing away with the less reliable chain.

THE PRINCIPAL PARTS OF A TRACTION ENGINE

A TYPICAL 6 N.H.P. AGRICULTURAL TRACTION ENGINE AS SUPPLIED BY JOHN FOWLER & CO. (LEEDS) LTD. ARRANGED TO SHOW SOME OF THE PRINCIPAL PARTS.

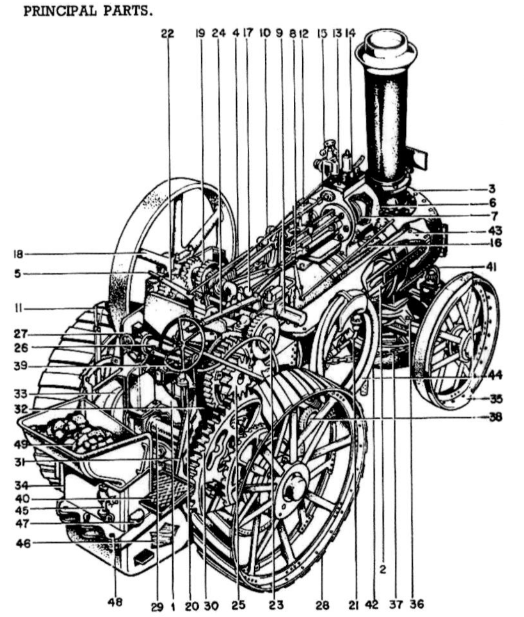

Key to numbered parts:

1. Firehole door.
2. Boiler barrel with fire-tubes inside, surrounded by water and leading to—
3. Smoke box.
4. Regulator rod, operated by the regulator handle (5) and so working the regulator valve itself which is situated in the top of the cylinder block, and which controls the flow of steam from the boiler to the valve chest.
6. Cylinder, with piston (7) inside, from which the piston rod passes out to the crosshead at the little end of the connecting rod (8).
9. The crankshaft is driven by the big end of the connecting rod, and carries two eccentrics to work the Stephenson's link motion (10) which works the valve inside the valve-chest.
11. The reversing lever.
12. Mechanical lubricator for pumping thick oil into the valve chest to lubricate the slide valve and the piston.
13. Whistle, (14) safety valve and (15) governors are mounted on top of the cylinder block, the latter being belt-driven from the crankshaft when in use to keep the engine running at a constant speed for threshing.
16. Cylinder drain cocks worked by the rod (17).
18. Flywheel.
19. Two-speed pinions. (N.B.—On many engines of other makes these are at the opposite end of the crankshaft.)
20. Feedpump to supply water to boiler.
21. Clackbox where water from pump enters the boiler. This contains a non-return valve and usually a shut-off cock to permit the boiler pressure to be shut off in the event of failure of the clack or any part of the piping.
22. Two-speed gears on second shaft (23) which slide on keyways to engage their respective partners on the crankshaft or lie disengaged between them according to the movement of the gear lever (24).
25. The third-shaft pinion (25) is keyed to the third-shaft (26) on which is a band-brake (27), although this is rather a feature of this type of Fowler and would not be found on other makes. Movement of the third-shaft is permitted by the slipper gear in the wheel No. (32), as the bearings of the third-shaft and the axle ("fourth-shaft") are connected and the two move together on the springs.
28. The offside hind wheel is driven directly by one pan-wheel of the differential, the other hind wheel being driven by the other pan-wheel via the axle (29).
30. The differential centre takes the drive from the third-shaft pinion (25) and transmits it to the two pan-wheels through the differential bevels.
31. Springs for hind axle and third-shaft.
32. Compensating gear.
33. Winding drum used when winching by withdrawing driving pins to leave near-side rear wheel free on axle.
34. Fairleads for winch rope.
35. Front wheels carried on swivelling forecarriage (36), controlled by steering chains (37) wound on steering drum worked by worm and worm-wheel (38) from steering wheel (39).
40. Steersman's platform, also forming footstep to footplate.
41. Oil lamps for night moves.
42. Spudpan intended to carry spuds (clips attached to rear wheels in muddy places) but usually used for other loose equipment.
43. Footboard. (44) Lift pipe used with water lifter (45) for filling tank (46), which can also be filled through the pocket (47).
48. Rear coupling with three pin-holes. (49) Bunker.

Fig 304: Traction Engine -Cutaway - Fowler

The fully-developed traction engine is typified by the example shown overleaf. This Marshall 7 nhp agricultural engine was built at their Britannia Works, Gainsborough, Lincolnshire in 1913. This is one of the most popular sizes of engine but they were also made in 5, 6, and 8 nhp sizes. Although rated at 7 nominal horsepower, the engine develops 24 brake horsepower. The motion runs at a governed speed of 160 rpm and at a boiler pressure of 100 psi, this is sufficient to power a full-size 54" thresher plus a chaff cutter and blower if so required. It weighs 10 tons empty and 11¼ tons in working order. At a maximum pressure of 140 psi, it can haul 36 tons on level ground and 20 tons up a gradient of 1 in 12. Most its working life would have been spent standing still so the springing to the rear axle used on road locomotives has not been incorporated. As well as reducing the initial cost of the engine, this has the added benefit that the engine can stand more firmly and not rock about. This was a particular advantage when driving a threshing machine, or indeed, any other static machinery such as stone-breakers or circular-saw benches.

Steering is effected by a swivelling simple beam front axle mounted below the perch bracket riveted to the underside of the smokebox. A chain fixed to each end of the axle passes around a 3" diameter barrel mounted in front of the firebox. The barrel is turned by a worm gear on the end of the steering column topped by a small hand wheel with a vertical handle.

Fig 305: Traction Engine - 7 nhp Marshall of 1913

This is very low-geared requiring about 9 turns from lock-to-lock. The rear wheel are of 6' 3" diameter x 16" wide and are constructed with cast-iron centres. The inner ends of the steel spokes are cast directly into the centres, the outer ends are splayed out and riveted to a pair of tee rings tied together with metal cross strakes which are also riveted on. These strakes are angled so as not to give a bumpy ride and set in such a direction that any side forces direct the wheels inwards towards the drive centres rather than against the retaining pins on the ends of the axles. To increase grip, a number of spuds or short angle brackets are carried and these can be bolted through holes in the strakes. Another option is to fit frost bolts that can be attached to the rear wheels in the same way. The front wheels are 4' 0" diameter x 9" wide and of similar construction to the rears but with a single tee ring and a plain iron tyre.

The double-acting single cylinder is 8½" diameter x 12" stroke and set on top of the boiler. The cylinder is enclosed in a steam jacket which also acts as a steam dome. This arrangement was patented by Thomas Aveling in 1861. The cylinder drives the crankshaft and its 4' 6" diameter x 6½" wide spoked flywheel. Final drive to the rear axle is via 2 intermediate shafts. 2 speed gearing gives maximum speeds of 1½ and 3 mph. Many engines were designed with a single intermediate shaft and the respective merits of 4-shaft and 3-shaft engines still fuel bitter arguments between the supporters of each system. All shafts are carried in hornplates. These are extensions of the firebox wrapper another idea originally patented by Thomas Aveling, this time in 1870. This engine like its cousin specifically designed for road haulage, the road locomotive, is fitted with a differential but most agricultural engines were not. The rear wheels are driven from the driving centres on the ends of the rear axle via pins that can be removed to disengage drive if so required when manoeuvring. Conversely, the drive pin can be replaced with a longer item inserted into the differential to lock the axle to prevent one wheel spinning.

The boiler is a cylinder of Siemens-Martin steel lagged with wood, and clad in light gauge steel sheet which is held in position by a series of brass straps. The boiler contains around 32 fire tubes of between 1¾" and 2½" diameter that run from the firebox at the rear to the smoke box at the front with the chimney above. The firebox is made up from wrought-iron plates and is double-skinned with water space between. Up to 1890 or so, all this would usually have been constructed from best quality wrought iron plate such as Staffordshire or Bowling and Low Moor from West Yorkshire. Later this changed to mild steel and Davey Paxman built one of the first portables to use a steel boiler in the early

1870's. Admission of steam to the cylinder is controlled by a disc-pattern regulator valve and a slide-pattern feed valve operated by link motion and controlled by the reversing lever. The most popular system for valve gear was stephenson link motion as in this example, but many alternative designs were offered, many based on the principle of a single eccentric as opposed to the stephenson's two. The huge range of these designs would fill a large book. The tender at the rear of the engine has enough room for 2 men to stand, one to drive and one to steer, plus 5 cwt of coal. Below the tender is a water tank holding 140 gallons of water. This is fed into the boiler via a non-return valve consisting of a brass ball resting on a seating in an enclosure known as a clack-box. When operating continuously, a pump driven from the crankshaft forces water into the boiler and the rate is controlled by adjusting the valve that controls the return water. On the move an alternative method is often used, this being the steam injector. Invented by aeronautical pioneer, Frenchman, Henri Giffard (1825-1882), this forces water into the boiler via a series of cones using the steam pressure of the boiler itself. On the face of it, this seems an impossibility and a bit like perpetual motion but work it does. What's more, it has no moving parts but the physics behind it is not easy to fathom. What all manufacturers stated was that any water used should be as clean as possible and in the USA, the advice given to drivers was "if you won't drink it, don't put it in your boiler".

One important feature of traction engines in general, rather, than the threshing engines in particular, is worthy of inclusion. That feature is compounding, sometimes known as expansive working. The idea is to take the exhaust steam from the one cylinder and allow it to expand again in a second larger cylinder. This principle was first applied to a traction engine by Fowlers of Leeds, and was demonstrated at the Royal Agricultural Society of England's show at Derby in 1881. Its improved efficiency and economy made it become the preferred design for road locomotives and was used virtually exclusively by the later steam tractor. What is not so well known is that Garretts had built a portable engine with compound cylinders the year before. Those in the threshing market generally did not consider that the reduction in running costs justified the additional initial outlay required to provide a much more expensive crankshaft assembly and an additional set of motion gear. One compromise solution was the tandem compound where the two cylinders were set end-to-end thus retaining the simplicity of the single motion gear. But none of these designs proved very successful, and certainly none were produced in numbers. The one compromise solution that was successful, and was used widely in the threshing trade, was Burrells's design for his "single-crank compound", the subject of patent no 3489 of 1889. Here the two cylinders were arranged in parallel, the high pressure cylinder diagonally above the low pressure cylinder, both linked to a common crosshead. At the Northampton Show of 1847, the twin cylinder engine exhibited by Mr Ogg of Northampton was roundly criticised by the judges. They expressed the opinion that "two cylinders are not required for agricultural purposes being more expensive to buy, with more parts to wear, and requiring more attention when operating them". Similar views were expressed when assessing threshing machines and remarks like "suitable for use by farmers rather than engineers" led to many designs that were crude in the extreme.

Fig 306: Traction Engine - 7 nhp Fowell of 1907

Fig 307: Export Strawburniner Marshall

Simplicity of design and cheapness were ideas that figured prominently in the design of machines intended for export. The finish was noticeably inferior, particularly the painting. Other features common on export machines were wider wheels to traverse softer ground and a flywheel clutch. A major feature of these engines was a variety of arrangements to burn inferior-quality fuels. All these machines had to be fitted with larger capacity fireboxes. If the engine was intended for straw-burning, particularly useful when driving threshing machines, the whole of the rear of the engine had to be open to allow sufficient access to the firebox, many being fitted with mechanical feeding apparatus. As a rule of thumb, it was considered that at least 5 times the weight of dry straw was required as against coal, and this could rise still further if the straw became wet. The chimneys of all strawburners were fitted with some form of spark arrester and many were fitted with auxiliary oil-burning apparatus. A Marshall catalogue mentions using 8 to 10 sheaves of straw to thresh 100 sheaves of corn. Straw as a fuel was never used in Britain. Mr Head of Ransomes, Sims, Head and Ransomes, Head & Jefferies, carried out a great deal of research on straw-burning engines around 1900 and was co-patentee of a feeder apparatus with one of their overseas agents, a Mr Schemioth. He made the following calculation: "An acre of straw weighs 30 cwt and is sold at 30/- per ton. Coal is priced at 20/- per ton. The calorific value of coal is 3½ that of straw, so the cost of straw as a fuel is over 5 times that of coal".

Table 9: Fuels for Steam Engines

COMPARATIVE EVAPORATIVE FUEL VALUES

Weight required to evaporate 8lbs of water

Fuel	**lbs**
Best Quality Welsh Steam Coal	0.53
Average Quality Coal	**1.01**
Dry Peat	2.01
Dry Wood	2.25
Dry Wood	2.31
Cotton Stalks	2.75
Brushwood	2.75
Straw	3.25
Straw	3.75

In the 1890's a number of engines were produced to a design to enabled them to be used as both general purpose traction engines and road rollers. The main reason for this was that the seasons for threshing and the rolling of water-bound macadam roads of the time were different. These were known as "convertible" engines and apart from the wheels and forecarriage/front roll assembly, differed little from their general purpose engine counterparts. Some of these engines were designed in such a way that they could also be converted to run as railway shunting engines.

The 1896 "Locomotives on Highways Act", better known as the Red Flag Act decreed that the previous draconian speed restrictions could be slightly eased in certain cases. Viz: A motor or light locomotive weighing less than 3 tons unladen, not drawing more than 1 vehicle, the unladen weight of both not to exceed 4 tons.
A few traction engine manufacturers took advantage of this change in the law to produce the first lightweight steam tractors. These were very small and commercially proved to be of very limited value. However with the principle established, there followed the 1903 "Motor Car Act". This amended the 1896 Act thus. Viz: A motor or light locomotive weighing less than 5 tons unladen, not drawing more than 1 vehicle, the unladen weight of both not to exceed 6½ tons. This size of traction engine was much more viable and many of the manufacturers of road locomotives went on to produce a large number of these

scaled-down machines. The 1927 "Heavy Motor Car (Amendment) Order" allowed the size to be increased for a second time to 7½ tons, but by this time the internal combustion engine was in the ascendancy and relatively few 7½ tonners were produced. Many of these later steam tractors went on to drive threshing machines, especially later on as the internal combustion engine's dominance on the road neared the complete.

Fig 308: Steam Tractor - 4nhp Garrett

Fig 309: Steam Tractor - 1929 Foden D-Type

A typical steam tractor is really a smaller version of the conventional traction engine, or more precisely, the road locomotive, but shares the majority of features with its agricultural engine counterpart. Rated at 4nhp and giving 20 bhp on a boiler pressure of 180 psi, at a governed speed of 230-250 rpm, it was just powerful enough to run a full-sized threshing machine, but would struggle with any additional load such as a chaff cutter for example. All of these engines are specifically designed so as to weigh less than 5 tons unladen as required under the 1903 Act. They have a small flywheel of 2' 9" to 3' 0" diameter x 5¼" wide. The flywheel is in the form of a solid disc and all the motion is enclosed in order to comply with a set of regulations intended to prevent the many horses, then on the roads, from being frightened. Two small compound cylinders of 5" and 8" bore x 8" stroke provide the power that is transmitted via 3 speed gearing and a differential. The 4' 9" diameter rear wheels are 12" wide and constructed with a single tee ring like the fronts which are 3' 1" diameter x 5" wide. Because the engine is designed to run on the road, all wheels are supplied standard with solid rubber tyres; two on each rear wheels size 1530 mm x 140 mm and one on each of the front wheels size 950 mm x 140 mm. Additional water is carried in a belly tank under the boiler like its road locomotive bigger brother and it can haul a load of 8 tons up a gradient of 1 in 12.

The steam tractor finally gave way to the steam wagon as the road transport vehicle of choice. By far the most successful builder was Fodens who produced over 7000. The earlier overtype design was also available in a steam tractor form known as the D-type and many of these went on to power threshing machines when changes in legislation made their use on the road no longer viable. Indeed some of the wagons were cut down as tractors and these too were used for threshing purposes.

In writing this book, the author is sometimes made to feel like someone writing the biography of someone whose brother was very famous and that the direction of the story is being inexorably led away from the original subject. The reader could also be forgiven for thinking that once the traction engine was perfected, the portable would be consigned to history, a view that would be supported by a visit to any modern steam rally. In fact nothing could be further from the truth. The market for portables was always much bigger than that for tractions and just as the first portable engines were built before the first traction engines, so the last portables were built after the last tractions and, indeed, the last threshing machines. To put this into perspective, Garrets of Leiston produced the largest number of steam tractors with 514 plus several hundred other traction engines. The number of portables they produced topped 33, 000 and Robey's were still building portables into the 1960's!

Chapter Seventeen

MOTIVE POWER 3 - AFTER STEAM

Lucas Headlamp - 1955 David Brown 25

The Internal Combustion Engine

The 1888 RASE Show held at Nottingham was the first offer of a medal for an oil engine. The first of a series of RASE Trials of petrol engines was held at Cambridge in 1894. Daniel Albone of Biggleswade, Bedfordshire, exhibited the first of his famous Ivel Tractors in 1902, and the RASE organised the first of their tractor trials near Baldock, Hertfordshire in 1910. At that event, the judges ruled that "steam had performed better on the day but that the future lay with the internal combustion engine. The First World War produced an enormous increase in tractor numbers. A total of 63,000 tractors were built in the United States in 1917, of which 15,000 were exported to Europe. In the following year, this total rose to 130,000. The Kent War Agriculture Committee reported that by the end of 1918, they alone were using 180 tractors. When the oil-engined tractor began to replace the traction engine, a belt pulley was often included in the basic design specifically to provide motive power for the threshing machine.

Fig 310: Albone's "Ivel" Tractor of 1902

BS 1495 of 1958 determined a number of design features for tractors including the position for the pulley, in front of the rear wheel and on the offside. This was driven off the engine's crankshaft via a bevel box at engine speed or slightly less. A belt speed of 3100 ft/min +/- 100 ft/min had been adopted as a standard in 1955, but the 8" diameter pulleys generally fitted as standard to the thresher required a belt speed of 2100 ft/min. This would mean fitting a larger replacement unit on the thresher. This would need to be in the order of 10" to 12" diameter.

Fig 311: Belt Pulley - 1955 David Brown 25

The two-speed power take off on the author's 1955 David Brown 25 tractor has a rear-mounted pulley 8½" diameter and 5¼" wide. British Standards of that time recommend a PTO speed of 536 rpm +/- 10 rpm. In that same year, the American Society of Agricultural Engineers suggested a second power take-off speed of 1000 rpm to complement the widely-adopted 540 rpm. At the preferred normal engine speed of 1300 rpm, the belt speeds for the high and low ratios on the pulley are 2100 and 3100 ft/min respectively.

Fig 312: 1942 Minneapolis Moline GTS

Fig 313: 1948 Fordson Major E27N P6

Just as the slower-turning portable engine carried a larger diameter flywheel than its traction engine successor, so the belt pulley of the slower-revving single cylinder diesel and semi-diesel tractors were of larger diameter than that of their multi-cylinder rivals. According to this logic it becomes possible to predict the optimum running speed of any engine intended to run a thresher simply by looking at the size of the flywheel or pulley.

Rather in the same way as operators of steam engines were advised to run their machines level so as to avoid the problems of priming if "head down" or uncovering the firebox crown if "head up", so operators of early internal-combustion engined tractors were advised to run their machines level so as to avoid the problems of oil starvation because of the simple splash lubrication systems used. It was also advised to earth tractors fitted with pneumatic tyres using a length of chain to prevent sparks caused by the build-up of static electricity when on the belt.

Fig 314: Field Marshall in need of restoration

Also to replicate the rope drum used on traction engines, many internal-combustion engined tractors used for threshing were fitted with a rear-mounted winch. Many of these featured a drop-down frame with ends to dig into the ground to give better purchase. As the older larger tractors were replaced by the new smaller machines such as the Ferguson TE20, their small size told against them when considering threshing. The author's 1955 David Brown 25 produces 30 bhp on TVO, easily powerful enough to drive a thresher; but weighing only 30 cwt, not nearly heavy enough to tow one safely.

Fig 315: Field Marshall with rear-mounted winch

THRESHER VERSUS COMBINE

Otherwise known as... where did it all go wrong?

1. The Thresher

... on the positive side...

A. Straw treated much more gently and left unbroken.
Better quality for resale - particularly when in sheaf form.
Suitable for: Thatching
Bedding
Potato clamps
Animal fodder

B. Weed seeds separated and retained for disposal / destruction rather than leaving them in the fields.

C. Seeds and Chobs discharged separately for recycling as poultry feed etc.

D. Better able to deal with heavily-weeded crops.

E. Better able to deal with very wet crops when used in conjunction with a binder.

... BUT...

A. Modern market for straw is minimal
Far less thatching
Far fewer horses

B. Heavily-weeded crops avoided by use of selective herbicides.
Especially juicy weeds such as thistles that can clog the screens and shakers of as modern combine.

C. Very wet crops avoided by hybridisation that provide early maturation thus bringing the time for harvest forward.

2. The Combine

A. Number of man-hours required FAR less.

B. Number of people required for seasonal work slashed.

C. Disruption to other farm work minimised

Chapter Eighteen

GOVERNORS

Wallis & Steevens Nameplate

In order for the threshing machine to run efficiently, the speed of the drum has to be kept constant. Despite the best efforts of a skilled feeder and team, the load on the source of motive power is bound to be constantly fluctuating. To achieve this end, all steam engines used for threshing were equipped with a governor. There were a number of different patented patterns but basically all worked in the same way. A number of spherical bob weights rotated about a vertical spindle mounted above the motion driven by a small belt off the crankshaft. As the speed of the spindle increased so centrifugal force moved the bob weights outwards. This movement was translated into a vertical movement by the spindle the bottom end of which acted on the valve admitting steam to the cylinder partially closing it. As the speed decreased so the valve spindle moved back towards its original position increasing the flow of steam.

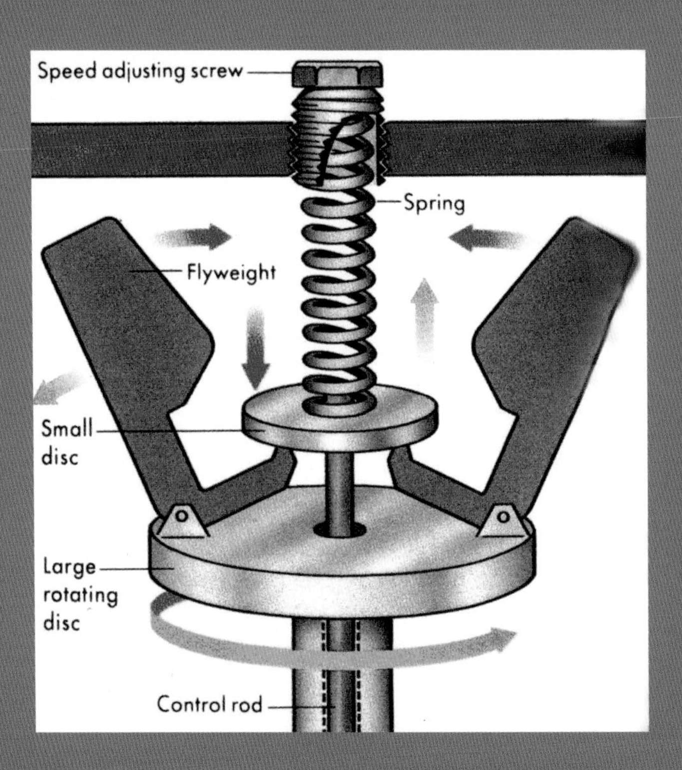

Fig 316: Diagrammatic Centrifugal Governor

Surprisingly, this design principle had been in use long before the advent of the steam engine. It had been used to control the speed of grindstones in mills by varying the pressure on the stones. This device mounted around the central spindle was curiously named a "lift-tenter".

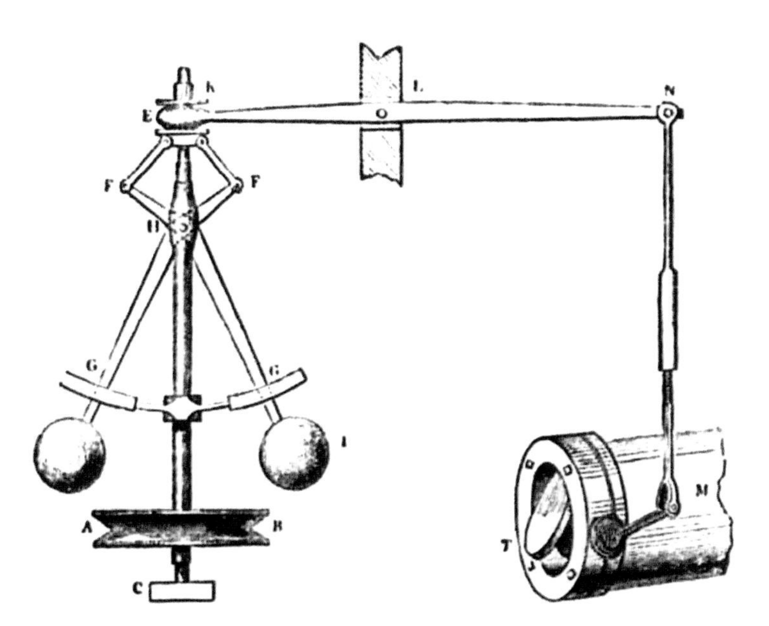

Fig 317: Early drawing of a Watt-type Governor

Early steam valves worked in the same way as the disc-shaped regulator valve. They partially rotated against a face machined flat, the disc gradually exposing a hole in the facing. This was a sound arrangement for the regulator but the friction of the disc on the face meant that it was heavy for the governor to operate. This resulted in the Watt type governor which was fitted with a large heavy pair of balls to overcome the resistance of the valve. The inertia of the weight of the large balls combined with the friction of the disc made these instruments unresponsive and they were sometimes described as being "slow-witted".

Fig 318: Watt-type Governor as fitted to Portable Engine

Chapter Nineteen

CONSERVATION & PRESERVATION RESTORATION & PRESERVATION

Rusty Nameplate

One of the main reasons for the disappearance of threshing machines is that they make too good a bonfire. Apart from the ease with which the wooden threshing machines can be set alight, they can also suffer from dry rot and woodworm.

The traction engine by comparison seems practically indestructible. Regularly a few examples are still retrieved from hedgerows or if abroad, the bush, and have even been known to have been exhumed from ponds. The thickness of their cast iron parts is their surest route to immortality.

Fig 322: Waiting for the Match

When considering the thresher in terms of re-use, if the innards are removed, then what is left is a box trailer, albeit rather large and cumbersome. If taken off its wheels, then it can become a shed, and some still exist in the Highlands of Scotland with thatched roofs masquerading as shepherd's huts. The innards themselves are of little value either as scrap or for recycling. But the good news is that threshing machines can still be used for part of their original purpose, namely the production of quality straw.

Rebuilding a thresher is less of a challenge than rebuilding a steam engine. Many of the parts are of wood and the metal components are relatively small. However the materials originally used are not as plentiful as they were. As "the Goons" used to say, "you can't get the wood..." This applies to both quality and quantity. Seasoned timber can be difficult if not impossible to source. Also large sections and long lengths can be difficult to obtain; boatbuilders can prove a useful source. On a smaller scale, if tongue and groove boarding is to be replaced, the size will be only slightly different. Visually this will not be noticeable but a small repair means exactly reproducing the original sections. This will require fabrication involving planing and routing. Some owners have tried to replace suspension slats with ply, but these tend to break because they lack the natural springyness of Ash.

Fig 323: Awaiting Exhumation

Fig 324: Ready for a Rebuild

Fig 325: In need of TLC

Fig 326: Broken Bottom Sill Member

Fig 327: Broken tongue and groove boarding

Straw walkers may look identical but many will not fit unless they are replaced in their original positions, and very probably in reverse order of how they were removed. Those replacing the brushes for rotary screens have had to resort to cannibalising brooms to obtain the bristles.

Although it is easy to bemoan the fact that seasoned timber is not readily available, some other modern materials are much better than their predecessors. Replacement drive belts feature a vulcanised joint that uses an 18" overlap, but these must be fitted so that they run in the right direction otherwise the joint can peel open. For use with a steam engine, these belts would generally be around 60' long, even longer when using a straw burner. If using a tractor, because of the smaller drive pulley, shorter overall length, and lower fire risk, this can be reduced to 50' or even 40'.

Graphite grease now often used for the lubrication of the fore carriage. While on the subject of lubrication, grease stauffers can usefully be replaced with conventional nipples provided that they are readily accessible. However this is not possible with larger capacity items. Trusser drive chains are often of a 52" pitch. A replacement unit from a modern combine harvester uses a slightly different construction but can be interchangeable.

A new sheet can be expensive but at least the modern materials used require no ongoing treatment. But some things don't change and generous applications of linseed oil are still thought to be the best treatment for ash hangers. If it takes the best part of a day to completely clean out a machine using a modern vacuum cleaner, how long must it have taken using just brushes? Or did they have the advantage of using a small boy for the more inaccessible areas rather in the manner of the stars of Charles Kingsley's "The Water Babies"?

Fig 328: Broken Pulley

Fig 329: Replacement of Shakers (Garvie)

The overriding consideration when demonstrating a threshing machine has to be safety. The dilemma is that the laissez-faire attitude of the Victorians and Edwardians is not shared by those in the 21st century. This problem is made worse by the fact that it is not just the safety of the operatives that has to be considered, but also that of the general public. If the machine is to be shown in operation, either at full speed as a threshing demonstration, or just turning over at an idle, it will require a separate power source. This may well mean a traction engine or portable and also means a long and exposed drive belt.

Fig 330: Plastic Net Fencing

Traditionally, the external belts on the machine itself would all have been uncovered which leads to a dilemma. On the one hand to stay "original" and leave everything unguarded or to cover up all moving parts. A compromise solution is to use weldmesh or similar material so that the moving parts can still be seen but not so easily touched but many find this unsightly.

At present, no additional guarding is required provided that a fence is erected a minimum of 9' away from the machinery. Owners are required to carry individual Public Liability Insurance.

At a rally, threshing demonstrations may be seen as presenting a lesser hazard as they are static. Powered vehicles circulating, albeit at relatively slow speeds present a far greater potential problem.

Also the threshing machine itself is potentially far less dangerous than a pressure vessel. Another advantage enjoyed by the thresher, and its cousin the reed comber, is that both are still in commercial use for the supply of thatching materials. This means that at least in principle, their operatives can use current skills. Conversely, the number of people who used steam traction commercially falls year on year, and in the second millennium, the number now must be tiny. A new generation of "preservation only" users cannot rely on the experience gained through normal work.

The National Traction Engine Trust was founded in 1954 and acts as an umbrella organisation for many other preservationist groups. It has been promoting its steam apprentice Club for over 30 years, and as a registered educational charity, this gives certificates of competency to its under-21 membership. Some feel that this system needs to be extended to all "modern" users of equipment at rallies. In parallel with examinations, it has been suggested that personal declarations of competence should be introduced, these being subject to some form of peer review.

In a world that is increasingly claim-conscious, rally organisers are anxious to ensure that they are seen to address all matters of safety, particularly those relating to the public. Being voluntary organisations, they cannot IMPOSE the statutory duties required by the 1974 Health & Safety at Work Act, but seek to apply its requirements in order to fulfil their "duty of care". This includes providing a Statement of Intent which "seeks to minimise the risk associated with the principal hazards associated with heritage plant. This is the notion of a "Safety Culture" described in the 1992 HSAW management regulations.

Fig 331: Sawbench guarded by pedestrian barriers

Fig 332: "A greater hazard"- a moving vehicle that is also a pressure vessel

Chapter Twenty

THE COMBINE HARVESTER METAMORPHOSIS

Headed paper from the Past

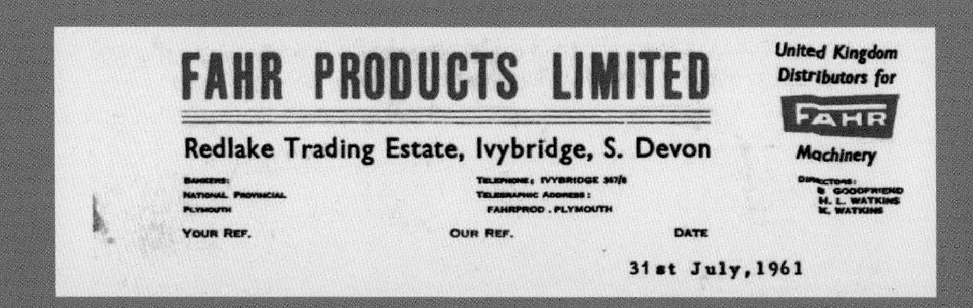

FAHR PRODUCTS LIMITED

Redlake Trading Estate, Ivybridge, S. Devon

United Kingdom Distributors for FAHR Machinery

BANKERS: NATIONAL PROVINCIAL PLYMOUTH

TELEPHONE: IVYBRIDGE 367/8
TELEGRAPHIC ADDRESS: FAHRPROD . PLYMOUTH

DIRECTORS: B. GOODFRIEND, H. L. WATKINS, K. WATKINS

YOUR REF. OUR REF. DATE

31st July,1961

An American article said "The separator (meaning the thresher) remained largely unchanged until the appearance of the combine harvester in the late 1920's, which could both cut and thresh at the same time. The combine did not, however, make the portable threshing machine obsolete and many remained in use through the 1940's".

Were this an English article, for 1920's and 1940's, it would read 1940's and 1950's.

The earliest record is of combine harvester is a machine built in Climax, Michigan USA, by Hiram Moore and John Hascall in 1834. It was pulled by a team of 20 horses and the mechanism was driven by a land wheel known in America as a bull wheel. Its 12' width of cut gave it a work rate of 10 acres per day. Over the next decade, various alterations to the basic machine made it more successful but few were sold. It was only in 1854 when a Moore-Hascall combine was shipped west to a dryer California that any real interest was shown. The new prototype succeeded in harvesting over 600 acres of wheat but unfortunately was destroyed by fire caused by an overheated bearing during the 1856 harvest. In spite of this setback, the potential of a combined harvester and thresher was realised, and between 1858 and 1888, at least 21 companies in the western states built combine harvesters.

Fig 333: 1884 horse-drawn Combine (Shippee)

In 1843 in Australia, prompted by an extreme labour shortage, a new design evolved based on the Roman vallus with a long fixed comb set at a height just below the ears of corn. The machine also featured a beater bar rotating above a concave. The crop was discharged into a box at the rear, but separate winnowing required. In 1885 Hugh McKay of Victoria incorporated the winnowing operation in his machine named the "Sunshine Harvester". The machine proved very successful and by 1900, many had been exported.

In the period 1875 to 1900, the major centre for combine harvester design was California. One of the reasons for this was, like Australia, the extremely dry operating conditions were ideal. However persuading the same design of machine to operate in a damper climate like Britain, proved much more challenging. Another aspect of harvesting that distinguished English practice was the value of straw. All American combine manufacturers were keen to encourage those using their machines to "take as little straw as possible". It was recognised that overall performance was limited by throughput of straw rather than grain. One way to achieve this was to keeping the cutter deck high and leave a long stubble.

A major step forward took place in 1871 when one B F Cook exhibited his combine at the California State Fair, held that year at Napa. Although it was still horse-drawn, the mechanism was powered by an oil-fired steam engine. 1886 saw the first self-propelling machine designed by George Stockton Berry. This used a single straw-fired boiler to power two separate sets of motion; one for the mechanism, and a second for propulsion. With a cut of no less than 40', it proved capable of a work rate of 100 acres a day. By 1890 2.5 million acres of wheat were being cut in California using combines, representing two thirds of the entire crop. The relative costs at the time were reckoned to be $3.00 per acre by a binder/thresher combination as against $1.50 for a combine. At that time, the two leading manufactures were both based in California, Holt Bros in San Leandro, and Daniel Best in Stockton. The two companies were finally to merge in 1925. The size of the turn-of-the-century machine had increased substantially and a report in a 1901 edition of "The Engineer" describes a combine in use in Washington State as having a 20' cut, weighing 8 tons, and being pulled by no fewer than 32 horses. John Deere started producing harvesting equipment in 1911, with their first combines being tested in 1925 and 1926. A batch of 40 prototypes were built in 1927 and full production followed in 1928.

Smaller lighter machines more suited to British conditions were being pioneered by the Massey-Harris company in Canada. Power take-off usually referred to simply as PTO had first been introduced in tractors in 1924 and this was used to power Massey-Harris's popular 'number 15' model of 1937. Their Clipper model of 1938 was the first to locate the thresher mechanism immediately behind cutter deck to give straight-through operation. As for their rivals, sales of the 1934 American Allis-Chalmers Model 60 with its 5' had risen by 1937 to 10,000 a year.

The first British design of combine harvester was by Clayton & Shuttleworth in 1928. This 12' cut machine was based very much on American lines using a small diameter pre-thresher drum and a peg-type main drum sometimes referred to as a "spike-tooth" drum. The beater bars on the drum were even described as "rasp bars", the term used on American machines. But it was the North American machines imported under the lend-lease scheme during WWII that started the great change from threshers. Numbers rose from 950 in 1942, to 2500 by 1944, and 10,000 by 1950. Some of these machines were assembled in Britain. Massey-Harris opened a factory in Kilmarnock, Ayrshire in 1949, followed by Allis Chalmers in Essendine, Lincolnshire, in 1950. A survey conducted in 1959 listed 26 machines by 12 different manufacturers, all but 7 being self-propelled. By 1960 the number of combines sold had topped 50,000.

As well as the machines imported from North America, a number were imported from Europe. Munktell of Eskilstuna, Sweden, had been a major producer of threshing machines since 1859 and nearly 13 000 were built before production ceased in 1951. Their prototype combine appeared in 1940, and their first production combine appeared 10 years later. Those imported into Britain were sold as A B Bolinder-Munktell up to 1973, when the name changed to B M Volvo. A prototype by Thermaenius followed Munktell in 1946, and Westerasmaskiner in 1948, who went on to export machines into Britain under the name Aktiv. As well as these other Swedish machines, the German company of Claas had produced their first machine, the "Mah-Dresh-Binder" in 1930, the first of these combines being imported into Britain in 1947. This was swiftly followed by Fahr which joined with Deutz in 1961.

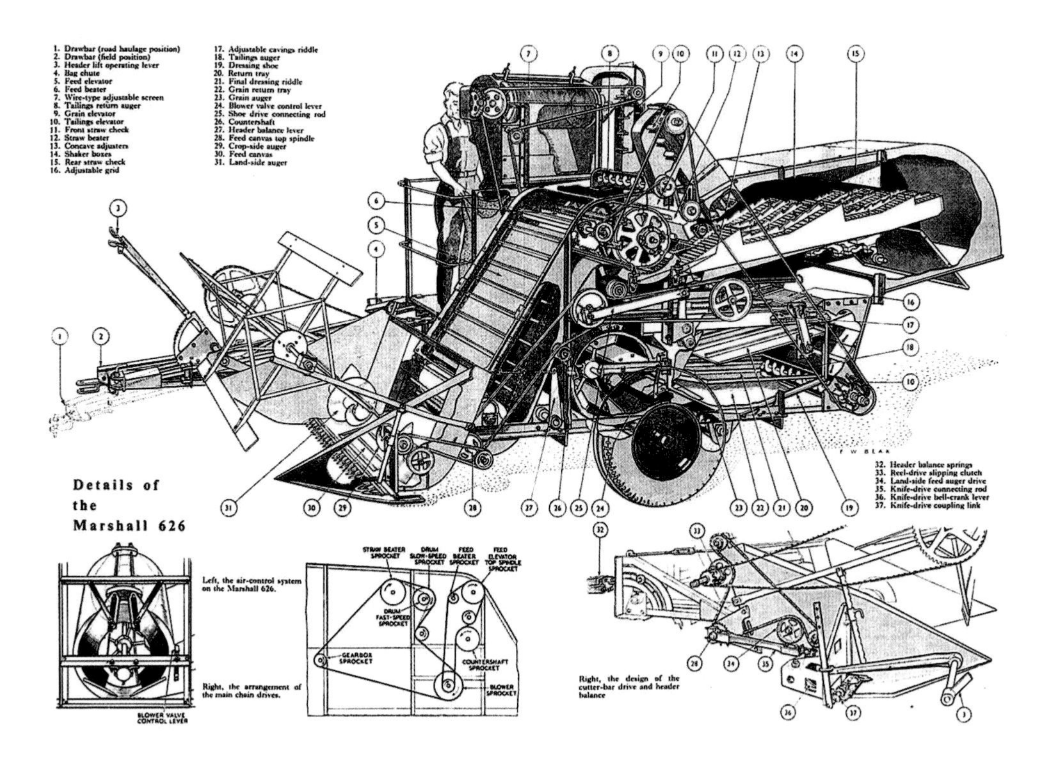

Fig 334: Early trailed-pattern Combine - Marshall 626

Many early trailed machines were powered by an auxiliary engine rather than by PTO, a similar arrangement being common on balers. A popular trailed machine was the Marshall 626. This had a cut of 5'9", later increased to 6'0". It was 22' long and weighed 2½ tons. A Ford 10 engine drove a 4 blade reel and side augers feeding a 3' wide canvas. The drum was 19" in diameter rotating over a 10 bar wired concave. Straw flow over a set of 4 shakers was controlled by 2 sets of check doors. The single 5 blade fan and the rotary screen were turned through 90 degrees like the first threshers to deliver grain to twin changeover best corn spouts and a single seconds/chaff/dust spout.

Unlike a thresher, the drum was fed by an additional feed drum of 9" diameter rotating at 660 rpm, and the straw was taken away by a matching straw drum also of 9" diameter and also rotating at 660 rpm. This gave a tip speed of these two units of 18 mph. The drum itself was 19" diameter and fitted with 2 drive pulleys. For working with grain, the smaller pulley was used giving a speed of 1320 rpm and a beater bar speed of 75 mph. For working with beans and other large crops, the larger pulley gave a speed of 725 rpm equivalent to a tip speed of 41 mph. The speed of delivery and removal via the feed and straw drums was 18 mph.

In the early 1940's, the first self-propelled machines began to appear. These were much more manoeuvrable in smaller fields with narrow gateways than their trailed predecessors and despite the high initial cost, soon consigned the trailed models to the history books. As with the thresher previously, the limiting factor was often its ability to fit down a narrow lane. A machine 7' 6" wide overall gave a 6' cut which meant a work rate of around 1 acre per hour.

A combine harvester could be seen as a binder with a threshing machine grafted on the back. But to enable it to perform effectively, it was necessary to make subtle alterations to the conventional thresher mechanism.

Fig 335: Early self-propelled-pattern Combine (1955 Fahr MD 5 A)

One important point has to be borne in mind when considering conventional combine harvester design; straw is a necessary evil whose impact has to be minimised, it has no value and can be treated with complete distain. Virtually no thatching meant virtually no market for long unbroken straw. In 1875, it was estimated that there were around 3.3 million horses in use in agriculture. Despite an upsurge in interest in shire horses, a survey conducted in 1986 could only identify 10,000. This represents a 97% drop over 100 years. A few horses as pets, rather than the major motive power means the market for any sort of straw as bedding is tiny compared with the past. And the same applies to chaff used for fodder.

This design approach is completely opposite to that of the thresher, where great time and trouble has been taken to produce a machine capable of delivering a quality straw sample. Because the straw is not required, horizontal rotors are often fitted to combines at the end of the shakers to spread the short lengths of straw allowing them to be ploughed in more easily.

In a combine harvester, the crop can be fed end-first, its passage through the machine can be via a number of secondary rollers and/or drums, and the main drum can be of the peg type. Any one of these arrangements is guaranteed to make the straw useless.

The two major design goals are to maximise throughput, and keep the machine reasonably compact. This is achieved in a number of ways.

Firstly there is no secondary treatment of the grain or dressing mechanism. In his book "Farm Machinery" written in 1963, Claude Culpin writes: "It is not usually possible to obtain a perfect sample with a combine harvester, neither is it desirable to overload the cleaning mechanism by attempting to do so. It is better to leave some of the cleaning to be done by barn machinery". The general quality of the output has improved considerably in the intervening period but even modern machines lack the sophistication of a thresher. Secondly, the shaker arrangement is abbreviated. Early Fahr machines used a single unit like the first Meikle patent. Often there is no separate grain board below the shakers, the grain being brought back to the middle of the machine via sloping trays integral with the shakers themselves.

Fig 337: Modern Combine - John Deere T Series

Just as the portable changed into the traction engine...

And the earliest threshers added winnowing and finishing...

So as we mourn the passing of the threshing machine, we can console ourselves with the knowledge that deep down inside every new shiny combine, there still lurks the drum, concave, walkers, shoes, and sieves of an old friend.

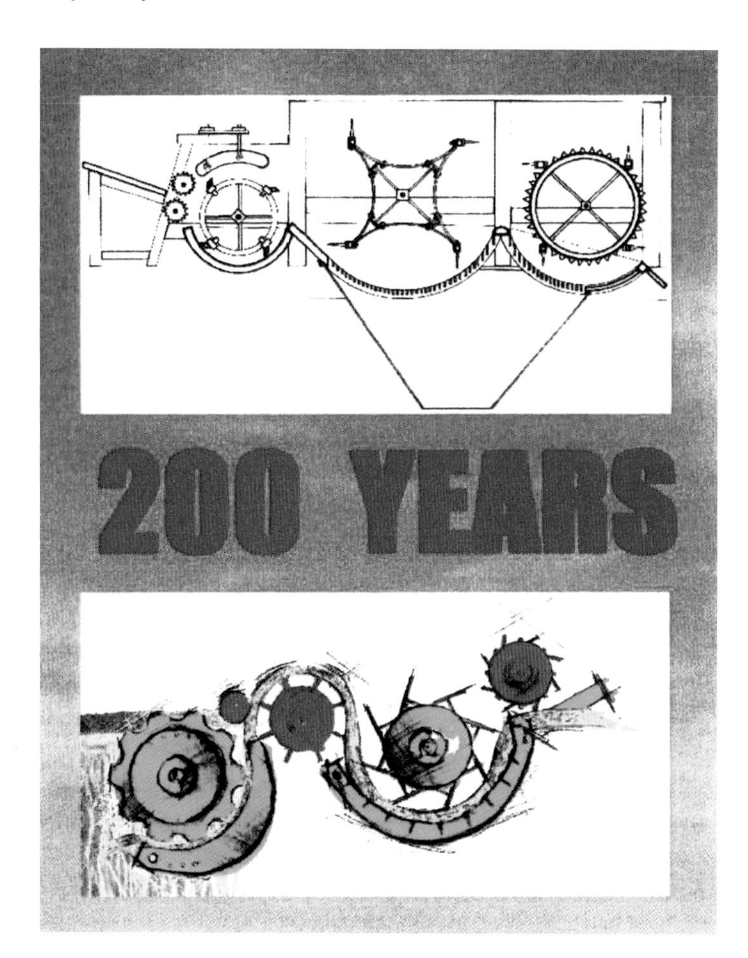

Frequently Asked Questions

Otherwise known as... the bluffers' guide to threshing machines.

Q. Why are threshing machines so large?
A. The greater distance the crop is allowed to travel, the greater the opportunity for the grain to separate from all the other material.

Q. Why are threshing machines generally red, orange, or pink?
A. The machines are largely made of wood. At the height of production, the most popular wood preservative material was red lead. The paints used contain a high lead content so tend to be a similar colour.

Q. What makes the modern machine the width it is?
A. The drum needs to be wide enough to allow the crop to be fed crosswise to minimise damage to straw and overall the machine must be narrow enough to fit down the narrowest of farm lanes.

Q. Why aren't threshing machines still in common use?
A. The machines require a team of operators the labour costs of which far outweigh those associated with the combine harvester.

Q. Is the combine harvester more efficient than the threshing machine?
A. No. The separation rates and quality of finished grain and straw are no better.

Q. Are threshing machines still in commercial use?
A. Yes, in very small numbers, but not for grain production. It is for the unbroken straw they produce that can be used for thatching.

Q. When were the first threshing machines built?
A. The first large static machines appeared just before 1800 in the Scottish Borders.

Q. When did the first modern pattern portable machines appear?
A. The design became standardised in around 1860 although simpler smaller machines did begin to appear in the 1840's.

Q. When was the zenith of threshing machine production?
A. The early years of the 20th century immediately before the first world war. Many of these were built for export.

Q. When did threshing machine production cease?
A. A few machines were built after the second world war with the very last appearing in 1961.

Q. What was the single most important developments in thresher design?
A. The "bolting" drum. So called because it was wide enough for bolts or sheaves to be fed crosswise allowing the straw to pass through unbroken.

Q. What was the single most important external influence on thresher design?
A. The availability of the small portable steam engine as a high-speed power source.

Q. If all this can be explained on a single page, why is this book so large?
A. Pass. Try reading James Joyce's "Ulysses"

Acknowledgements

Trelowarren, Mawgan-in-Meneage, Cornwall. c1910

The author would have been totally unable to complete this work without the valuable assistance of those listed below that he CAN remember... and lots of others that he CAN'T. To those he offers his sincerest apologies.

The Farmers

Reg Bennett	Kehelland
Lewis Hosking	Treswithian
William Olds	Nancemellin

The Preservationists

Willy Alford	Threshers
Colin Benney	Threshers
Bob Franklyn	Barn Thresher
Brian Johns	Thresher
Roger Knowles	Nameplates
Peter Lobb	Thresher
Henry & Jane Marks	Combine
Maurice Raby	Thresher
Paul Richards	Tractor
John Thomas	Thresher
Mike Wevell	Baler
Dick Wood	Reed Comber

The Specialists

Paul & Jane Cockerham	Equestrian
Ken Hall	Agricultural Engineering
Richard Olds	Agricultural Contracting
Charlie Osborne	Seeds & Fertilizers
Mike Pawluk	Thatching

Illustrations

Burrell Wheel Centre

Terminology conventions

Fowler Wheel Centre

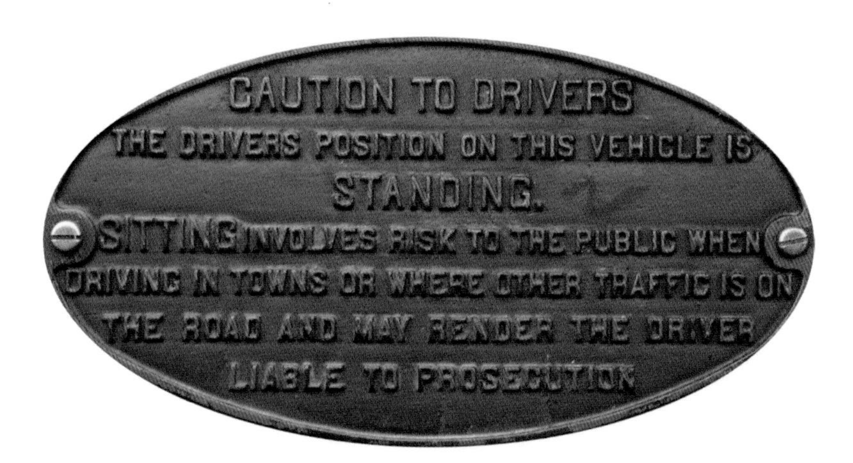

In compiling his book, one quickly becomes aware that everyone seems to have their own name for everything one is trying to describe. This variety of terms applies not only to users of the machines, but equally to their manufacturers. This could result in a substantial part of the text taken up with phrases like "also known as" and "often referred to as" etc. To save print, generally one term will be used throughout. Below is a list of just some of the other terms used with their equivalent used in the text. At this point, the author is painfully aware of Ralph Waldo Emerson's immortal words:

"Consistency is the hobgoblin of small minds"

Bottom Blower =	Main Fan
Bottom Shoe =	Main Shoe
Box =	Thresher
Cavings Sieve =	Cavings Riddle
Corn and Chaff Receiving Board =	Top Shoe
Drum =	Threshing drum within the Concave
First Fan =	Main Fan
Harps =	Shakers
Hummeller =	Awner
Jog Crank =	Riddle Crank
Machine =	Thresher
Mill =	Thresher
Piler =	The complete Awner / Chobber / Smutter Unit
Riddle Box =	Dressing Shoe
Riddles =	Coarse Sieves
Second Fan =	Dressing Fan
Shoe Crank =	Riddle Crank
Smutter =	Chobber
Thrasher =	Thresher
Top Blower =	Dressing Fan
Vibrators =	Shakers
Walkers =	Shakers

GLOSSARY

Clayton & Shuttleworth Thresher Wheel Hub

Otherwise known as…

"An expedition to the outermost regions of the Anglo-Saxon language…"

Awn Fibrous spike attached to the sheath of a grain of barley or wheat.
Cape Separated ear of grain still enclosed by its husk
Cavings Small items other than grain separated from the long straw.
Chaff Small lengths of cut straw
Chaff Short cut lengths of straw
Chob Separated ear of grain still enclosed by its husk
Creeper Auger
Fanning To blow away lighter material leaving grain
Felloe Section of the rim of a wooden wheel
Gaitin Single sheaf fanned out at base and stood on end
Haver Grain used as fodder….as in sack
Heckle To beat flax
Hulling Breaking the heads of grasses in order to remove the seeds
Hummelling The removal of awns
Nave The hub of a wheel
Polishing The removal of husks from grain
Riddle Sieve often to remove coarser material
Scutching Drawing flax through a comb to remove the seeds
Sheaf A bundle of corn, straw, or hay usually tied
Spikelet Awn or other similar small growth
Stook A group of 6 to 12 sheaves piled together so as to aid drying
Winnowing To blow away lighter material leaving grain
Wort Malted grain such as barley used for beer and whiskey-making

IMPERIAL UNITS AND CONVERSIONS

Makers Plate - 1050 rpm

Length
18 inches = 1 cubit
11 cubits = 1 ox goad, rod, pole, perch
4 ox goads, rods, poles, perches = 1 chain
12 inches = 1 foot
3 feet = 1 yard
6 feet = 1 fathom
22 yards = 1 chain
10 chains = 1 furlong
8 furlongs = 1 mile

1 inch = 25.4 millimetres
1 foot = 305 millimetres
1 yard = 914 millimetres
1 mile = 1.61 kilometres

Area
272.25 square feet = 1 perch
151.55 perches = 1 acre
4640 square yards = 1 acre
640 acres = 1 square mile

1 acre = 0.405 hectares
1 square mile = 2.56 square kilometres

Pressure
1 pound per square inch = 6.89 pascals

Sacks
Size = 2' 3" x 4' 4" weighing 4 lbs

Volume
4 gills =1 pint
2 pints = 1 quart
4 quarts = 1 gallon
8 gallons = 1 peck
4 pecks = 1 bushel
1 bushel = 1.28 cubic feet.
4 bushels = 1 coomb
4 bushels = 1 sack
6 bushels = 1 boll
2 sacks = 1quarter
1 Hungarian mandelu = 15 sheaves

1 imperial gallon = 4.54 litres

Weight
16 ounces = 1 pound
14 pounds = 1 stone
8 stones = 1 hundredweight
20 hundredweights = 1 ton

1 pound = 0.454 kilograms
1 hundredweight = 50.85 kilograms
1 ton = 1.02 metric tonnes

BIBLIOGRAPHY

Guernsey Guille-Alles Library Label "Books must not be entrusted to Children, nor must they be exposed to Rain."

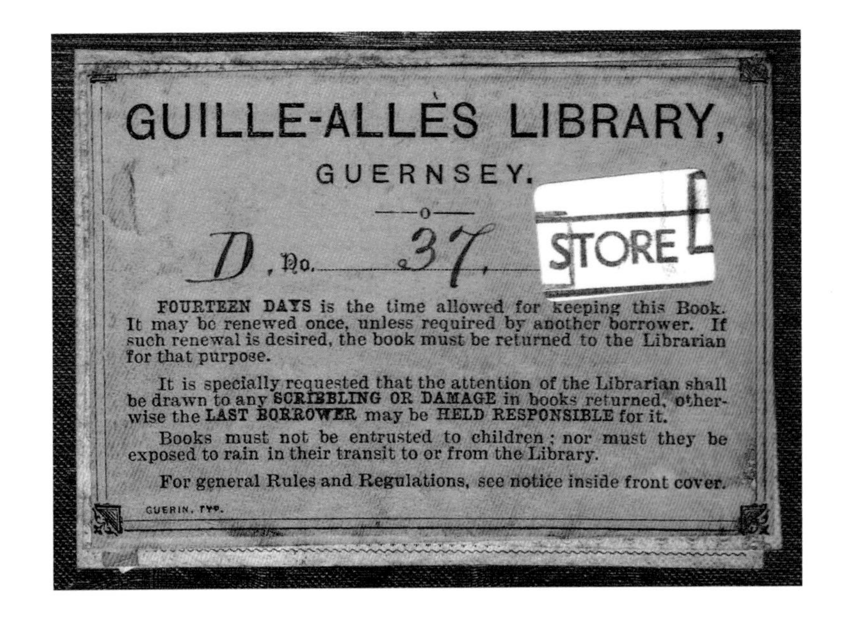

GUILLE-ALLÈS LIBRARY,

GUERNSEY.

—o—

D, No. 37. STORE

FOURTEEN DAYS is the time allowed for keeping this Book. It may be renewed once, unless required by another borrower. If such renewal is desired, the book must be returned to the Librarian for that purpose.

It is specially requested that the attention of the Librarian shall be drawn to any SCRIBBLING OR DAMAGE in books returned, otherwise the LAST BORROWER may be HELD RESPONSIBLE for it.

Books must not be entrusted to children; nor must they be exposed to rain in their transit to or from the Library.

For general Rules and Regulations, see notice inside front cover.

GUERIN, TYP.

"If you copy from one source it's called plagiarism, if you copy from many it's called research."

Books

The Implements of Agriculture	J A Ransome	J Ridgeway	1843
Rudimentary Treatise on Agricultural Engineering	G H Andrews	John Weale	1852
The Book of the Farm - 3rd Edition	H Stephens		1876
Field Implements & Machines	J Scott	Crosby Lockwood	1884
Barn Implements & Machines	J Scott	Crosby Lockwood	1904
The Steam Portable Engine	W D Wansbrough		1912
Steam Engine Builders of Essex, Suffolk & Cambs	R H Clark Goose		1950
Mechanised farming Esso			1950
Traction Engines worth modelling	W J Hughes	David & Charles	1950
Cereal Varieties in Great Britain	R A Peachey	Crosby Lockwood	1951
Chronicles of a Country Works	R H Clark	Percival Marshall	1952
Technical & Economics of the Combine (Rpt 81)	A H Gill	Bristol University	1955
Steam Engine Builders of Lincolnshire R. H. Clark			1955
Know your Tractor - a Shell Guide Shell Petroleum			1955
A Century of Traction Engines	W J Hughes	Percival Marshall	1959
Traction Engines P Wright A & C Black			1959
The development of the English Traction Engine	R H Clark Goose		1960
Farm Machinery	C Culpin	Crosby Lockwood	1963
Garretts of Leiston	R A Whitehead	MAP	1964
Modern Manual for drivers of Steam Road Vehicles	W M Salmon	MAP	1967
Taskers of Andover (Waterloo Iron Works)	L T C Rolt	David & Charles	1969
Victorian Engineering	LTC Rolt	Allen Lane	1970
Farming with Steam	H Bonnett	Shire	1974
Traction Engines	P Wright	A & C Black	1959
The Complete Traction Engineman	E E Kimbell	Ian Allan	1973
An Illustrated History of Traction Engines	P Wilkes	Spur	1974
Farm Tractors in colour	M Williams	Blandford Press	1974
Ransomes of Ipswich	D R Grace	Inst of Agric Hist	1975
Victorian Farming - A Source Book	A Jewell	Redwood Burn	1975

Title	Author	Publisher	Year
Steam Power in Agriculture	M Williams	Blandford Press	1977
The Foden Story	P Kennett	Patrick Stephens	1978
The Grain Harvesters	G Quick / W Buchele	ASAE	1978
Introducing Model Traction Engine Construction	J Haining	Argus Books	1983
Wallis & Stevens - A History	R A Whitehead	Road Loco Soc	1983
Portable Steam Engines	L R Shearman	Shire Publs	1986
Massey-Ferguson Tractors	M Williams	Blandford Press	1987
Harvests of Change - history of RASE	N Goddard	Quiller Press	1988
Steam Engine Builders of Norfolk	R H Clark	Haynes Publishing	1988
Harvesting Machinery	R Brigden	Shire Publs	1989
Wellington Ironworks (Fosters of Lincoln)	M R Lane	Quiller Press	1990
The Story of Britannia Iron Works (Marshalls)	M R Lane	Quiller Press	1993
Threshers Yard	J L Compton	EATEC	1993
Crop Production Evolution, History & Technology	C W Smith	John Wiley	1995
Steam Thrashing 1900 - 1950	G F A Gilbert	G F A Gilbert	1996
The British World Book Encyclopedia	BWEInternational		1997
Great Harvests with Bolinder-Munktell	C J Huggett	Gote House Publ	1999
A Brief History of Twine	D Lipski	Madison Art	2000
The Countryside Remembered	S Ward	TAJ Books	2004
Heavy Horses	D Zeuner	Shire Publs	2004
The Magic of Old Tractors	I M Johnston	New Holland	2004
Tractors - Historical view of the modern tractor	M de Cet	Abbeydale	2007

Articles etc

Title	Author	Publisher	Year
Application of steam power to agricultural purposes	William Waller	Inst Mech Eng	1881
On Threshing Machines	W W Beaumont	Inst Mech Eng	1881
Applications of engineering to agriculture	F Ayton	Inst Mech Eng	1926
The Origins of the Threshing Machine	W Tritton	Lincolnshire Mag	1934

Clayton & Shuttleworth brochure	1910
Foster catalogue	1959
Fahr Combine Harvester handbook	1961
Claas Combine Harvester workshop manual	1970
John Deere brochure	2007
Cornish Studies Library Archive	2008

What OTHER people have said...

The Threshing Machine was very highly thought of in its day.

"Threshing by steam was the most notable early application of power to agriculture, and the threshing machine evidences a higher character of mechanical arrangement than attaches to any other implement of agriculture."
James Allen Ransome - Ipswich - 1843

"This is thus a somewhat complicated piece of combined mechanism, which is purely the result of experiment, unaided by any application or theory to the phenomena or laws upon which the successful performance of its various and complex functions depends".
William Worby Beaumont - London - April 1881

"There is perhaps no single machine designed to carry out any series of operations or processes, in which so many conditions or circumstances are involved, and have to be fully considered and provided for, as in the finishing thrashing machine".
Prof. John Scott - London 1904

"With the exception perhaps of the reaper and binder, the most interesting and at the same time complex machine which the farmer uses is the thrasher."
Frank Ayton - Ipswich - June 1926

Much less highly thought of were some of its Operators.

"Farmers are generally ignorant of the principles of Mechanics".
G. H. Andrews - London - 1852

"Beware of drivers of the semi-semi-skilled type"
Arthur Wedgwood

And always, there is the whiff of steam...

"Threshing without steam is like eating bread without butter"
Gilmar Johnson, Frederic, Wisconsin